NEW WARS AND NEW SOLDIERS

It is supremely difficult to "teach" ethics—and equally important to do so. These chapters are accessible, diverse, intriguing. On one level they show us troublesome issues of the globe's newest contest grounds peopled with mercenaries, illegals, terrorists, and innocent civilians. At their best, these chapters help us with what contributor Rebecca Johnson thoughtfully calls "moral capacity building for irregular warfare".

Christopher C. Harmon, Marine Corps University, USA and author of *Terrorism Today*

Tripodi and Wolfendale have brought together a multidisciplinary team of authors to address a wide range of ethical dilemmas faced by soldiers in the new missions conducted in unconventional battlefields. Readable, wide ranging and engaging, this book analyses the ways in which the principles of military ethics can be applied to contemporary warfare and reminds us that this subject should not be neglected in the education of today's soldiers.

Christopher Dandeker, King's College London, UK

Aimed at both military and academic audiences, the book strikes a nice balance between scholarly rigour and practical relevance. Its chapters are grouped into several main themes: discussions of the continued relevance of 'just war theory'; the implicit and explicit obligations of nations that engage in humanitarian intervention; the ethical implications of new technologies and new actors in the battlespace; and finally a 'rubber-hits-the-road' examination of the implications of all this for combat behaviour, leadership and training ... thoughtful and provocative ... I commend this book to professional readers at all levels, and particularly senior leaders.

Australian Defence Force Journal

Warfare has undergone a radical transformation in recent decades. The objectives of war; the agents of war; the targets and victims of war; the methods and weapons of war; and the nature of the battlefield have all changed. New Wars and New Soldiers *not only subjects these various developments to much-needed ethical scrutiny; it also explores their implications for the normative frameworks that have traditionally been applied in evaluating recourse to war, conduct in war, and post-war relations ... This short review cannot do justice to the impressive breadth of Tripodi and Wolfendale's collection ... The volume features ... useful insights and recommendations regarding the education and training of troops. What is more impressive is that this scope does not come at the expense of depth and detail. Tripodi and Wolfendale have produced one of the most enlightening, dynamic, and cutting-edge collections on military ethics in recent memory.*

Journal of Battlefield Technology

This valuable set of chapters will benefit anyone perplexed about the global war on terrorism, the inequality of military power, and the moral principles of just war theory and preventive wars ... Highly recommended for lower-level undergraduates through professionals/practitioners; general readers.

Choice

This book brings together experts on military ethics to contribute to this debate and to discuss the new challenges of the twenty-first century that have arisen that question the validity of the just war tradition in an era of globalization and the changing nature of the military's role ... Overall this is a very good book. It is thought provoking and challenging, which should lead to further debate on the subject. The book is easily accessible to the reader, as it is well written with concise chapters. This book should be of interest to the military, security personnel, academics, and students studying in the fields of international relations, security studies, politics, history, sociology, and philosophy. I recommend this book.

Democracy and Security

This book seeks to interpret ethical risks, in contemporary times, for one of humankind's oldest and most destructive endeavours: warfare. What is being placed under the microscope is the seemingly oxymoronic phenomenon of military ethics, which the editors define as a discipline covering many issues, 'from states' rights, to when and how military force may be deployed, to whether or not military forces can ever intentionally cause the deaths of civilians'. All these critical issues are robustly debated in this important book.

New Zealand International Review

New Wars and New Soldiers
Military Ethics in the Contemporary World

Edited by

PAOLO TRIPODI
Marine Corps University, USA

JESSICA WOLFENDALE
West Virginia University, USA

ASHGATE

Published by
Ashgate Publishing Limited
Wey Court East
Union Road
Farnham
Surrey, GU9 7PT
England

Ashgate Publishing Company
Suite 420
101 Cherry Street
Burlington
VT 05401-4405
USA

www.ashgate.com

British Library Cataloguing in Publication Data
New wars and new soldiers : military ethics in the
  contemporary world. -- (Military and defence ethics)
  1. Military ethics.
  I. Series II. Tripodi, Paolo. III. Wolfendale, Jessica,
  1973-
  172.4'2-dc22

Library of Congress Cataloging-in-Publication Data
New wars and new soldiers : military ethics in the contemporary world / by Paolo Tripodi and Jessica Wolfendale.
    p. cm. -- (Military & international relations)
  Includes index.
  ISBN 978-1-4094-0105-6 (hardcover) -- ISBN 978-1-4094-0106-3 (ebook)
  1. Military ethics. 2. Just war doctrine. 3. Military psychology. 4.
  Humanitarian intervention--Moral and ethical aspects. 5. Military art and science--
Technological innovations--Moral and ethical aspects. 6. Terrorism--Prevention--
Moral and ethical aspects. 7. Private military companies--Moral and ethical aspects.
8. Private security services--Moral and ethical aspects. I. Tripodi, Paolo, 1965- II.
Wolfendale, Jessica, 1973-
  U22.N48 2011
  172'.42--dc22

                                                                    2010048087

ISBN 978 1 4094 0105 6 (hbk)
ISBN 978 1 4094 5347 5 (pbk)
ISBN 978 1 4094 0106 3 (ebk)

MIX
Paper from
responsible sources
FSC® C018575

Printed and bound in Great Britain by the
MPG Books Group, UK.

Contents

PART IV   NEW ACTORS IN THE BATTLEFIELD

PART V   COMBAT BEHAVIOR AND TRAINING

List of Contributors

Deane-Peter Baker is Assistant Professor of Philosophy in the Department of Leadership, Ethics and Law at the United States Naval Academy. Before taking up his current position in January 2010, Dr. Baker was Associate Professor of Ethics in the School of Philosophy and Ethics at the University of KwaZulu-Natal in South Africa, where he taught for eleven years. Dr. Baker is a member of the U.S. Naval Academy's Africa Forum and Irregular Warfare Working Group, and was formerly Director of the University of KwaZulu-Natal Strategic Studies Group. Dr. Baker retired from the South African Army Reserve in 2009 with the rank of Major. He served as Convenor of the South African Army Future Vision Research Group and as part of the leader group of the Umvoti Mounted Rifles, a reserve armoured cavalry regiment. A specialist in both epistemology and the ethics of armed conflict, Dr. Baker's research straddles philosophy, ethics and security studies. From 2006 to 2008 Dr. Baker was Chairman of the Ethics Society of South Africa. He has held visiting fellowships at the Triangle Institute for Security Studies in North Carolina and the Strategic Studies Institute of the U.S. Army War College. Among his numerous publications is *Just Warriors, Inc: The Ethics of Privatized Force* (Continuum, 2011).

Peter Bradley is a former lieutenant-colonel in the Canadian Forces. He teaches Military Psychology and Ethics at the Royal Military College of Canada.

C.A.J. (Tony) Coady was formerly Boyce Gibson Professor of Philosophy at the University of Melbourne and is now Professorial Fellow in Applied Philosophy at the Centre for Applied Philosophy and Public Ethics (CAPPE) in that university. He is best known for his work in epistemology, especially his book *Testimony: A Philosophical Study* (Oxford University Press, 1992), and in applied philosophy where he has written extensively on issues to do with war and terrorism, much of which is further developed in his book *Morality and Political Violence* (Cambridge University Press, 2008). Several of his papers on the definition and morality of terrorism have been anthologized. In 2005, he gave the Uehiro Lectures on Practical Ethics at Oxford University. An expanded version has been published as the book *Messy Morality: The Challenge of Politics* (Oxford University Press, 2008).

Stephen Coleman is Senior Lecturer in Ethics and Leadership and Vincent Fairfax Foundation Fellow in the School of Humanities and Social Sciences, at the University of New South Wales at the Australian Defence Force Academy. He is the author of *The Ethics of Artificial Uteruses: Implications for Reproduction* *and Abortion* (Ashgate, 2004) and many papers in academic journals and edited collections on a diverse range of topics in applied ethics, including military ethics, police ethics, medical ethics, and the practical applications of human rights. In addition to these published papers he has presented at conferences in Australia, New Zealand, Hong Kong, the United Kingdom and the United States. At ADFA he teaches courses in military ethics and practical ethics at both undergraduate and postgraduate level.

Nick Evans is a Ph.D. candidate at the Centre for Applied Philosophy and Public Ethics, at the Australian National University, Canberra. Formerly a graduate of the University of Sydney in Physics, his interests are the social and moral responsibilities of scientists, just war theory, and the history and philosophy of science. His dissertation, due for competition in 2010, is on the dual-use dilemma in the life sciences.

Rebecca Johnson is an Assistant Professor of National Security Affairs at the Command and Staff College at Marine Corps University where she teaches courses in Culture, Interagency Cooperation, and Military Ethics. Before coming to Marine Corps University, Dr. Johnson taught courses on international ethics, religion and conflict, and post-conflict reconstruction at Georgetown University. In 2006 she was selected to participate in the United States Institute of Peace's faculty seminar on "Global Peace and Security from Multiple Perspectives." Previously, she was the Academic Director for the South Carolina Washington Fellows Program at the University of South Carolina. She is a member of the Inter-University Seminar on Armed Forces and Society, International Studies Association, Women in International Security, and attends the annual meeting of the International Symposium of Military Ethics. Her most recent publication, "Fight Right to Fight Well: General McChrystal's Approach to Preserving Noncombatant Immunity" appeared in the April 2010 edition of *Small Wars Journal*. Dr. Johnson serves as the book review editor for the *Journal of Military Ethics*.

David Lefkowitz is an Assistant Professor in the Philosophy department at the University of North Carolina–Greensboro, and currently a Laurance S. Rockefeller Visiting Research Fellow at Princeton University's Center for Human Values. His research focuses on the morality of obedience and disobedience to law, philosophical issues in international law, and questions of war and morality. Publications in the last of these areas include a chapter on collateral damage in *War: Essays in Political Philosophy* (Cambridge University Press, 2008); "Partiality and Weighing Harm to Non-Combatants" in the *Journal of Moral Philosophy* 6:3 (2009); and "On Moral Arguments Against a Legal Right to Humanitarian Intervention," in *Public Affairs Quarterly*, 20:2 (2006).

George R. Lucas, Jr. is Professor of Philosophy and Director of Navy and National Programs in the Vice Admiral James B. Stockdale Center for Ethical

Leadership at the United States Naval Academy (Annapolis, MD), and Visiting Professor of Ethics at the Naval Postgraduate School (Monterey, CA). He has taught at Emory University, Georgetown University, and served as Chair of the Department of Philosophy at Santa Clara University, and also as Assistant Director in the Division of Research Programs in the National Endowment for the Humanities. He was a Fulbright Fellow in Belgium in 1989, and held a fellowship from the American Council of Learned Societies in 1983. Lucas received his Ph.D. in Philosophy from Northwestern University in 1978. He is author of four books, over fifty journal articles, translations, and book reviews, and has also edited eight book-length collections of articles in philosophy and ethics, including most recently: *Anthropologists in Arms: The Ethics of Military Anthropology* (AltaMira Press, 2009), *Ethics and the Military Profession: The Moral Foundations of Leadership* (New York: Pearson Education, 2005), *Perspectives on Humanitarian Military Intervention (*University of California Press, 2001), "From *Jus ad bellum* to *Jus ad pacem*: Rethinking Just War Criteria for the Use of Military Force for Humanitarian Ends," in *Ethics and Foreign Intervention* (Cambridge University Press, 2003), and "The Role of the International Community in the Just War Tradition: Confronting the Challenges of Humanitarian Intervention and Preemptive War," *Journal of Military Ethics* (2003). He is co-editor of the new international book series, "Defence Ethics" (Farnham: Ashgate), and Program Chair of the International Society for Military Ethics.

Peter Olsthoorn holds a Ph.D. degree from Leiden University in political theory, and teaches at the Netherlands Defence Academy. He has written several articles on military ethics, and the book *Military Ethics and Virtues: An Interdisciplinary Approach for the 21st Century* (Routledge, 2010).

Gerhard Øverland is Project Leader, Centre for the Study of Mind in Nature (CSMN), University of Oslo and Senior Research Fellow, Centre for Applied Philosophy and Public Ethics (CAPPE), Charles Sturt University. He specializes in moral and political philosophy and the ethics of war. His work has been published in the *Journal of Moral Philosophy*, *Public Affairs Quarterly, Ethics, and Ethics & International Affairs.*

Robert Sparrow is a Senior Lecturer in the School of Philosophy and Bioethics at Monash University in Australia. He has published widely in applied ethics and political philosophy. His current research interests include the ethics of military robotics, just war theory, and the ethics of nanotechnology. He is the author of "Predators or Plowshares? Time to Consider Arms Control of Robotic Weapons," *IEEE Technology and Society*, 28:1: 25–9 (2009); "Killer Robots," *Journal of Applied Philosophy*, 24:1: 62–77 (2007); and "'Hands Up Who Wants To Die?': Primoratz on Responsibility and Civilian Immunity in Wartime," *Ethical Theory and Moral Practice*, 8:3: 299–319 (2005). He is currently working on an Australian Research Council-funded Discovery Project, "Good Soldiers and Ethical Soldiers," with Dr. Jessica Wolfendale and Professor Tony Coady.

Uwe Steinhoff is Assistant Professor at the Department of Politics and Public Administration of the University of Hong Kong and Senior Research Associate in the Oxford Leverhulme Programme on the Changing Character of War. He is the author of *The Philosophy of Jürgen Habermas: A Critical Introduction* (Oxford University Press, 2009), *On the Ethics of War and Terrorism* (Oxford University Press, 2007), *Effiziente Ethik: Über Rationalität, Selbstformung, Politik und Postmoderne* (Mentis, 2006) and of numerous articles on war, terrorism, ethics and political philosophy.

Paolo Tripodi is Professor of Ethics and Ethics Branch Head at the Lejeune Leadership Institute, Marine Corps University, USA. He served as Professor of Strategic Studies at the Marine Corps War College. He was the MCU Donald Bren Chair of Ethics and Leadership. He has held academic positions at the U.S. Naval Academy in Annapolis, MD; the Catholic University of Chile, Santiago, Chile and Nottingham Trent University, Nottingham, UK. Dr. Tripodi trained as an infantry officer and received his commission with the Italian Carabinieri upon graduation from the School of Infantry and Cavalry in Cesano, and the Carabinieri Officers' School in Rome, Italy. Dr. Tripodi is the author of *The Colonial Legacy in Somalia* (Macmillan 1999) and of more than twenty refereed articles some of which were published in *The Journal of Military Ethics*, *International Journal on World Peace*, *Small Wars and Insurgencies*, *Security Dialogue*, *International Peacekeeping*, *Low Intensity Conflict and Law Enforcement*, *Medicine, Conflict and Survival*, *International Relations*, and the *Journal of Strategic Studies*.

Jessica Wolfendale is an Assistant Professor in the Philosophy department at West Virginia University, Morgantown. Prior to taking up that position in July 2009, she was an Australian Research Council Postdoctoral Research Fellow at the Centre for Applied Philosophy and Public Ethics at the University of Melbourne. She is the author of *Torture and the Military Profession* (Palgrave Macmillan, 2007), and has published extensively on topics including torture, military professional ethics, and terrorism.

List of Acronyms

AAA	American Anthropological Association
AMRs	Autonomous Military Robots
ANC	African National Congress
AWS	Autonomous Weapon Systems
BCI	Brain–Computer Interface
BCT	Brigade combat team(s)
CAOCL	Center for Advanced Operational and Cultural Learning
CAR	Canadian Airborne Regiment
CEAUSSIC	Commission on the Engagement of Anthropology with the U.S. Security and Intelligence Community
CoE	Code of Ethics
COIN	Counter Insurgency
CORDS	Civil Operations Revolutionary Development Support
DARPA	Defence Advanced Research Projects Agency
DFAC	Dining facility (or "mess hall")
DND	Department of National Defence (Canada)
DOD	U.S. Department of Defense
ECOMOG	Economic Community of West African States Monitoring Group
ECOWAS	Economic Community of West African States
EMP	Electromagnetic Pulse
EO	Executive Outcomes (a private military company)
EU	European Union
FOB	Forward Operating Base
FRAGOS	Fragment Orders
HANDS	Human Assisted Neural Devices
HTS	Human Terrain System
HTT	Human terrain team(s)
HUMINT	Human intelligence (i.e., gathering information via direct human contacts)
ICI	Oregon – International Charter Incorporated of Oregon (a private military company)
ICISS	International Commission on Intervention and State Sovereignty
IED	Improvised explosive device
ISAF	International Security Assistance Force
IW	Irregular warfare

JAB	Jus Ad Bellum
JEB	Jus Ex Bello
JIB	Jus In Bello
JPB	Jus Post Bellum
JWT-I	Irregular just war theory
JWT-R	Regular just war theory
KKK	Ku Klux Klan
LZ	Landing Zone
MCIA	Marine Corps Intelligence Activity
MHAT	Mental Health Assessment Team
MI	Military Intelligence
NATO	North Atlantic Treaty Organization
NCO	Non-commissioned officer
NGO	Non-Governmental Organization
OAU	Organization of African Unity
OPORDS	Operational Orders
PAE	Pacific Architects and Engineers (a private military company)
PFA	People's Front of Ardania (fictional organization)
PMC	Private Military Company/Contractor(s)
PMF	Private Military Firm
PoW	Prisoner of War
PRT	Provisional reconstruction teams
PTSD	Post Traumatic Stress Disorder
RCT	Regional combat team(s)
RMA	Revolution in military affairs
RtoP	Responsibility to Protect doctrine
RUF	Revolutionary United Front (Sierra Leone)
SME	Subject matter expert
TRADOC	Training and Doctrinal Command, U.S. Army
UAV	Uninhabited/Unmanned Aerial Vehicle
UCAV	Uninhabited Combat Aerial Vehicle
UMS	Unmanned Systems
UN	United Nations
URV	Unmanned Robotic Vehicle
VBIED	Vehicle-Borne Improvised Explosive Device

Acknowledgments

This book emerged out of a workshop funded by the Australian Research Council and GovNet on the topic of "New Wars and New Soldiers", held at the Centre for Applied Philosophy and Public Ethics at the University of Melbourne on August 11–13, 2008. This workshop brought together academics, military officers, and members of the general public to discuss ethical issues in modern combat. The excellent papers presented by the participants of that workshop form the basis of several of the chapters in this volume. Additional chapters were invited from prominent authors in military ethics to complete the fascinating and thought-provoking collection contained in this book.

This book could not have been completed without the support of the Australian Research Council, or without the enthusiasm and encouragement of the Ashgate Military and Defence Ethics series editors George Lucas, Don Carrick, James Connelly, and Paul Robinson. From 2008 to 2009, Jessica Wolfendale's work on this volume was supported by the Australian Research Council as part of a Postdoctoral Research Fellowship. Paolo Tripodi was the MCU Donald Bren Chair of Ethics and Leadership from 2003 to 2009. During those years, his research was supported by the Marine Corps University Foundation. MCUF also provided support for this book. Paolo Tripodi would also like to thank Andrea Hamlen for her advice and suggestions on his text.

We would also like to express our deep gratitude to our contributors for the interesting, challenging, and thoughtful chapters they have written. Their willingness to engage wholeheartedly with the many difficult and controversial issues in modern military ethics (and responding to our feedback and comments) is inspiring, and we are extremely impressed with the quality of the contributions to this book, and with the depth and thoughtfulness they display.

Introduction

Paolo Tripodi and Jessica Wolfendale

The Development of Modern Military Ethics

Military ethics is a discipline that covers a wide range of issues, from states' rights, to when and how military force may be deployed, to whether or not military forces can ever intentionally cause the deaths of civilians. Traditionally, however, military ethics has paid little attention to those individuals in uniform whose main objective, if and when they are deployed, is to kill, and yet it is this that makes military ethics such a fascinating and a challenging topic. For many years, it was widely assumed that soldiers would learn to fight honorably according to a usually unspoken and unofficial code of ethics or a code of chivalry, yet training soldiers in ethical behavior was rarely considered an essential part of military culture. War is brutal, after all, and so it seems inevitable that those involved in it will adapt to a certain degree of brutality. Violence managed by individuals in uniform had and continues to have one main objective: victory. The defeat of the enemy, with a minimal loss of friendly casualties, has constituted the main focus of the analysis of the military historian and political analyst. How wars have been fought and who has fought them, has frequently been relegated to a merely secondary interest.

Thankfully, in the early 1990s this way of analyzing war began to change. This change did not simply reflect the end of the Cold War and thus the end of the prospect of a conventional war terminating in a nuclear holocaust. Instead, this new way of thinking about military ethics emerged in concert with the increasing use of military forces in operations that did not resemble conventional battlefields. The use of military force for humanitarian and peacekeeping operations began to raise questions about soldiers' ability to accept these new roles and their readiness to perform them. The Canadian Airborne Regiment's abuse and killing of innocent civilians during the 1992 intervention in Somalia (Winslow 1997) and the many failures of that mission (Thakur 1994; Tripodi 1999; Clarke and Herbst 1996) suggested that soldiers whose mind-set and training was mostly geared to conventional warfighting had serious difficulties in adapting and performing such a mission while remaining within the ethical constraints of war. This new phenomenon attracted the interest of sociologists (Moskos and Miller 1995) as well as psychologists (see Littz et al., 1997), who explored the complexities of peace operations for individual soldiers and the friction soldiers experienced when deployed in "unfamiliar" environments.

Many of the issues addressed by scholars of these disciplines had implications for applied ethics. How should the military employ force in environments that were significantly different from traditional battlefields—environments no longer characterized by a simple friend-enemy dichotomy but complicated by large numbers of civilians whose status as friend or foe, innocent or non-innocent, could be very unclear? The 1992 mission in Somalia ended in a failure that had disastrous consequences for Somalia and the Somali people. The debate that followed Somalia focused mostly on whether soldiers should or should not perform these types of missions.

While the debate about Somalia was ongoing, a UN mission deployed in Rwanda under the command of Canadian general Roméo Dallaire was unable to stop one of the most violence genocides of the 20th century. In just one hundred days close to a million people were murdered, in many cases sadistically and brutally. At the beginning of the genocide, ten Belgian paratroopers were lynched by a mob of rogue Rwandan Government soldiers. This event influenced the decision of many of the military contingents deployed in Rwanda to expose themselves to as little risk as possible (Tripodi 2006). As a result, in several instances these contingents failed to protect defenseless civilians (Gourevitch 1998). The situation in Rwanda dramatically illustrated the ethical question of whether soldiers should be willing to accept a significant risk of death in order to protect and defend individuals with whom they might have nothing in common and for whose sake they were acting.

What brought the awful challenges soldiers operating in a "peace operation" faced to the attention of the western world was the 1995 massacre in Srebrenica. In the small enclave of Srebrenica, an isolated battalion of Dutch soldiers faced an overwhelming force of determined Bosnian Serbs who were intending to massacre the male Muslims who had sought shelter in the city. The Dutch battalion was not only operating below full strength, running low in supplies of ammunition, food and gasoline, and in a poor strategic position, but morale was also extremely low. Many of the Dutch soldiers began to believe that they had been given a "mission impossible"; that they had been forgotten by the higher chain of command, and were now forced to protect people with whom they had little connection and even less understanding (Honig and Both 1996).

The slaughter of approximately 8,000 people in the Balkans, so close to Western Europe, finally pushed the international community to take a more assertive stance on the issue of humanitarian intervention. This more assertive approach proved long-lasting, and in 1999 world leaders, including Tony Blair and Bill Clinton, were able to convince NATO leaders that, if the murder of ethnic Albanians in the Serbian province of Kosovo could not be halted, war would be a necessary course of action. In March 1999, NATO forces launched an air campaign against the Serbian government of Milošević. The war over Kosovo sparked a new debate about the use of force in international relations and whether the international community had the right to interfere within the borders of a state if that state's government victimizes citizens. Similarly, NATO's decision to permit pilots to fly at higher altitudes in order to avoid anti-aircraft fire, thereby lowering the accuracy of their bombing raids and so increasing the risk to civilians, provoked a related debate about the relative trade-off between risk to intervening soldiers, and the risk to civilians posed by the use of military force in interventions.

The 1990s aroused the interest of scholars in the nature, conduct, and moral dilemmas of soldiers' new missions, as well as the role of the international community in protecting defenseless victims of violence even when those victims were located beyond state national borders.

The debate about humanitarian and peacekeeping missions has continued into the 2000s, but military ethics has also expanded to explore issues such as the threat of international terrorism, the development of autonomous weapons systems, and the rapid rise of private military and security companies. The ongoing wars in Iraq and Afghanistan have crystallized many of these issues. The Bush administration attempted to justify the 2003 invasion of Iraq by claiming that a war in order to prevent a future but not imminent threat could be permissible—a claim that extends the traditional justification for war well beyond defensive and preemptive war, as it is traditionally understood. Both the Iraq and Afghanistan wars also involved the extensive use of private military and security companies, raising questions about the accountability of these companies, particularly in relation to war crimes. This concern became particularly pertinent after the killing of seventeen civilians by personnel from the private military company Blackwater—none of whom were convicted. The use of unmanned aerial vehicles has also been prevalent in both combat arenas, presenting us with the unprecedented spectacle of operators in Seattle piloting planes thousands of miles away. Last, but far from least, the events at Abu Ghraib and Guantánamo Bay have raised serious questions about the use of torture in the fight against terrorism, as well as the efficacy of the current ethics training of military personnel.

In response to these issues, military ethics gained particular importance and prominence during the 2000s. In 2001 the *Journal of Military Ethics* was launched, a journal that has since become increasingly central to debates about ethics and the military. The *Joint Service Conference on Professional Ethics* (renamed the *International Symposium on Military Ethics* in 2005) began to attract growing interest from scholars and practitioners in the US and abroad throughout this decade, and comparable initiatives were launched in the UK and in Europe, notably the *Military Ethics Education Network* established in 2008 at the University of Hull, UK.

Military Ethics and the Challenges of Modern Conflict

The above developments demonstrate that military ethics today is a thriving field of study. Yet, military ethics continues to be in a state of evolution. In 2010 the editors of the *Journal of Military Ethics*, Martin Cook and Henrik Syse (2010, 119), suggested that "military ethics is a species of the genus 'professional ethics'" that is "analogous to medical ethics or legal ethics." This conception of military ethics captures the way in which military ethics has developed as a field encompassing a wide array of ethical issues from codes of ethics for professional soldiers to the ethics of preventive war, as well as the growing interdisciplinary nature of military ethics today.

Military ethics now includes work from disciplines as diverse as philosophy, history, sociology, international relations, politics, and psychology. The increasing diversity of approaches to military ethics suggests that scholars of military ethics will be better prepared for their analysis if they are open minded and ready to explore areas whose relevance to the field might not be immediately clear. The approach to contemporary military ethics that is best suited to understanding soldiers' roles in the wars of today is therefore a broad, multidisciplinary vision.

The contributors to this volume take such an approach, and the chapters in this book clearly demonstrate a desire to contribute new knowledge and new reasoning to the field of military ethics. Several chapters examine the practical issues facing new soldiers in new forms of combat, while other chapters approach modern warfare from the perspective of theoretical philosophical analysis. Philosophers and other academics are sometimes criticized for not engaging with the reality of combat or the nature of the modern world. Yet it is a serious mistake to dismiss philosophical analysis on these grounds. Philosophical and theoretical analysis is crucial to military ethics, since too often developments in the conduct and resort to war outstrip careful analysis of the ethical issues raised by those developments. Many debates about modern combat rely on assumptions about, for instance, the duties of soldiers to undertake risks to save civilians, that are rarely clearly articulated, let alone argued for. Theoretical analysis is thus essential to a thorough understanding of the ethics of modern combat. The combination of theoretical, empirical, and practical approaches in this volume thereby provide a nuanced view of when and how new forms of combat present new ethical dilemmas, and how these dilemmas might be resolved.

New Wars and New Soldiers

This book addresses key issues in the nature of modern combat, the role and training of soldiers in new conflict environments, and the ethical issues raised by the presence of new actors on the battlefield, such as the involvement of military anthropologists in the wars in Iraq and Afghanistan and the use of private military companies. This comprehensive approach allows the reader to see the broad scope of modern military ethics, and to understand the numerous questions about modern conflict that require critical scrutiny.

Part I addresses the relevance of just war theory to modern combat. In light of the events of the first years of this century, it is essential to consider whether just war theory can provide the normative framework through which modern conflicts—including modern terrorism—may be assessed, or whether the threat of modern non-state terrorism and other modern wars renders the moral constraints of just war theory obsolete and even dangerous.

Since 9/11 many politicians, academics, and commentators have argued that the threat of terrorism means that abiding by the constraints of just war theory will be ineffective at best and suicidal at worst. Jessica Wolfendale considers this argument broadly—stepping back from the specific issue of terrorism to ask whether (and when) new aspects of war constitute a reason to revise or loosen the constraints of just war theory. She argues that those aspects of modern warfare that have been claimed to be new have in fact been features of warfare throughout the 20th century. Instead, the new aspects of modern war are the creation of a global war on terrorism and the extreme imbalance in military power engendered by the rise of the US military force. Yet even these new elements do not provide strong reasons to rethink just war theory. Instead, the basic moral principles that underlie just war theory provide important guidelines and a fundamental moral framework that can be brought to bear on modern warfare.

Stephen Coleman responds directly to a recent book by Nick Fotion, in which Fotion argues that the jus ad bellum criteria of just cause and last resort should be weakened for states fighting non-state actors, and furthermore that states (but not non-state forces) should be permitted greater latitude in the application of the principle of discrimination in such conflicts. Coleman carefully considers and rejects Fotion's arguments for these revisions to just war theory, and concludes that even in conflicts in which one side uses terrorism, traditional just war theory can be applied to both sides. Furthermore, Coleman suggests, non-military ways of fighting terrorism have been unfortunately neglected in the literature on this topic, yet a legal response could be both more effective and less destructive than a military response.

C.A.J. Coady discusses a recent addition to just war theory—the theory of jus post bellum, or justice in the aftermath of war, which encompasses moral considerations about the process of ending wars as well as moral considerations governing post-war relationships between the victorious and the defeated. For Coady, the wars in Iraq and Afghanistan offer important lessons in the failure of the aggressive imposition of external reconstruction, especially when motivated by a sense of moral righteousness. Combined with a careful discussion of the duties that a just victor might have toward a defeated enemy, Coady offers strong arguments against aggressive and punitive post-war policies that severely damage the possibility that the defeated state can regain political autonomy and that may unfairly punish civilians, who bear little if no responsibility for the war.

Part II turns to the question of humanitarian intervention. For many years it was widely believed that states had a right to non-interference which imposed a duty on outside parties not to intervene in a state's internal affairs, even if that state was committing human rights abuses against its own citizens. However, since the 1980s, this belief in the right to non-interference has come under increasing fire. Now, with the development of the Responsibility to Protect doctrine (ICISS 2001) among the international community, there is growing acceptance of the view that a state's right to non-interference depends on whether that state protects its citizens' basic human rights. If a state fails to do so or deliberately violates those rights, then external military force may be justified to intervene to protect that state's citizens.

However, the theoretical underpinnings of humanitarian intervention have been poorly explored. It is not clear, for example, how much risk soldiers of the intervening state should be willing to shoulder in order to protect the civilians on whose behalf they are intervening. Should states be willing to take on significant levels of risk to save the lives of civilians, even at substantial cost to soldiers' lives? In practice, the answer tends to be "no", as was clear from the position adopted by NATO during the civil war in the former Yugoslavia, when NATO forces decided to bomb Kosovo from a higher altitude, which protected pilots from anti-aircraft fire, but increased the risk of civilian casualties. Similarly, it is unclear whether states might have a moral duty to intervene to prevent gross human rights abuses, and if such a duty exists, how it should be enforced. These issues indicate how important a clear theoretical discussion of humanitarian intervention is. Gerhard Øverland and David Lefkowitz provide such an analysis in this volume.

Øverland argues that, rather then thinking of humanitarian assistance as a moral duty (as many commentators have argued), humanitarian assistance is derived from the principle of assistance—a principle that imposes weaker demands on the intervening party than the more stringent idea of a duty to intervene. This analysis leads Øverland to conclude that since the civilians are the beneficiaries of intervention (because they suffer so severely under their current government's regime), they do not have a right to insist that intervening soldiers bear significant risks to rescue them. Thus, soldiers do not have a moral duty to shoulder considerable risk to reduce harm to civilians, particularly since the civilians themselves would (we can assume) be willing to shoulder a degree of risk to bring an end to their suffering at the hands of their current leaders.

Unlike Øverland, David Lefkowitz argues that there is a duty to intervene that arises from a duty to take on a fair share of burdens in order to ensure the equal protection and respect of basic human rights, and the subjects of this duty must be willing to accept certain restrictions on their freedom as well as certain levels of risk in order to carry out this duty. Thus, Lefkowitz argues, in situations where there are insufficient volunteers to conduct a humanitarian intervention, a process of conscription could be justified in order to fairly distribute the burdens of undertaking justified interventions. He concludes that in cases where a proposed humanitarian intervention meets the requirements of the jus ad bellum, the costs to individuals in waging such an intervention are not unreasonable.

In Part III, Nicholas Evans and Robert Sparrow explore the ethical implications of modern weapons development—specifically, the development of what is known as "neurowarfare," and the increasing use of robotics and autonomous weapons systems on the battlefield. The challenge when examining the ethical issues posed by these developments is to distinguish whether and in what way this weaponry raises genuinely new ethical issues. How are these technologies different, ethically speaking, from the invention of the bow and arrow and the machine gun?

Nicholas Evans discusses a truly unique development in war—Human Assisted Neural Devices (HANDs). HANDs allow soldiers to control military equipment through a neural connection, thus allowing them to literally control the equipment with their minds. While HANDS raise ethical issues similar to those associated with other long-range and autonomous weapons systems (Killmister 2008), Evans argues that HANDs raise the possibility of truly risk-free warfare and thus could create a deep asymmetry between forces that possess this technology and those that do not. The ethical concern with such asymmetry, according to Evans, is that military personnel who conduct war via HANDs may not have moral permission to kill their enemies, since the enemy they fight against has no feasible way of fighting back and so, to all intents and purposes, is not a combatant in any meaningful sense. Thus the technology that makes war easier and safer to perform may at the same time undermine the moral justification for the use of lethal force in war—a consequence that has serious and widespread implications for just war theory.

Robert Sparrow explores the ethical issues raised by the ever more sophisticated development of military robotics. For Sparrow, the use of unmanned weapons systems and autonomous military robots raises ethical concerns that pose significant challenges for just war theory. For instance, the widespread use of unmanned systems affects considerations of proportionality and discrimination by increasing the possibly of accurate targeting—thus it could become unethical to use less sophisticated weapons that lack such accurate targeting systems. Similarly, if military robots become capable of making autonomous decisions involving the use of lethal force, there are clear questions about accountability and responsibility, as well as questions about whether robots could learn to distinguish combatants from noncombatants, and possess the right attitude to those they kill in combat.

Part IV turns to the question of new actors in the battlefield. While mercenaries have been involved in warfare since war began, the corporate private military company is a relatively new development—and one that challenges the traditional view of mercenaries as immoral profit-hungry killers. In their chapters, Uwe Steinhoff and Deane-Peter Baker study the role of private military/security companies (PMCs) in modern conflicts. Steinhoff focuses on the role played by private security companies in Africa, while Baker makes extensive use of Peter Feaver's Agency Theory in order to analyze the appropriate role of PMCs more broadly. Steinhoff and Baker both argue against the traditional rejection of mercenaries while recognizing that the involvement of private military contractors raises serious issues in relation to accountability and regulation. Yet, Steinhoff suggests that from a moral perspective the disadvantages of mercenarism do not outweigh the advantages. Thus what is required is not the abolishment but the regulation of PMCs. This would result in a significant reduction of the disadvantages currently present in the use of PMCs, while increasing the advantages that PMCs offer. Baker contrasts the role of PMCs with that of state military forces, in particular looks at how the concept of honor affects members and leaders of both state military and private military companies. From a moral point of view he suggests that both organizations have a role to play in different operational settings. For instance, state military forces might be better suited for fighting wars of national emergency, since they are trained to accept an ethics of sacrifice, while PMCs might be more advantageous in operations that are "in support of humanity."

The presence of anthropologists on the battlefield has generated a lively debate that has become particularly intense among anthropologists themselves. George Lucas, the author of an important book on this topic (Lucas 2009), explores the role that anthropologists have played in helping the military develop a better understanding of what is known as "Human Terrain." Lucas considers two extreme positions inside the profession—those anthropologists who believe that the greater involvement of anthropologists in the HTS (Human Terrain System) will make a positive impact by saving innocent lives and reducing casualties; and those who firmly believe that any involvement of anthropologists in military operations is both unethical and will have a negative outcome. Lucas suggests that in many respects the presence of anthropologists within military units and their role in raising troops' cultural awareness raises similar ethical issues as the practice of having chaplains and medical personnel embedded with the troops. The risk posed by the presence of such individuals must be carefully balanced against the benefits of such involvement.

In the final part of the book, the focus turns to the ethical challenges faced by individual soldiers in the battlefield, specifically in counterinsurgency (COIN) environments. Paolo Tripodi analyzes how such environments can compromise soldiers' ability to take the morally correct course of action. Tripodi explores how the physical conditions and psychological dimension of the battlefield environment can have a serious emotional impact on soldiers. He investigates how soldiers operating in such an environment might step in what he calls the "evil zone," a space in which even individuals equipped with the best character might perpetrate atrocities. Therefore, he concludes, it is important that soldiers possess a high level of "situational awareness" and, even more importantly, that they are aware of the powerful force that situations can exert on moral decision-making.

Peter Bradley explains how situational forces and several other key factors, above all poor leadership, led to the infamous events perpetrated by members of the Canadian Airborne Regiment (CAR) in Somalia in the early 1990s. Using James Rest's four psychological components for proper moral functioning, Bradley investigates the determinants of moral behavior: individual factors and situational influences. His analysis focuses on how military culture and the military mindset can have a detrimental impact on soldiers' moral decision making and may lead to unethical behavior. Bradley emphasizes the importance of developing training programs that would enhance soldiers' moral awareness and judgment. In his view, had the CAR soldiers gone through the training programs he suggested, they would have been aware of environmental and other potentially negative factors, such as the influence of peer and group behavior. This training may have better prepared them to make morally sound decisions.

In her chapter, Rebecca Johnson deals with the complexities of irregular warfare and the moral challenges that soldiers face when deployed in what General Krulak defined as the "three-block war." In particular, Johnson focuses on the strategic corporal and the role he/she plays as a small unit leader. She assesses the importance of unit cohesion as a key element of an effective fighting team, and investigates the elements of moral behavior and places them in the context of irregular warfare. She provides an enlightening assessment of how James Rest's four elements for moral behavior—moral sensitivity, moral judgment, moral motivation, and moral character—apply to Army and Marine Corps character development programs. She emphasizes that all four elements are crucial for developing soldiers' and Marines' ability to act morally in the highly stressful and demanding environment that is characteristic of irregular warfare.

Peter Olsthoorn examines and seriously questions the idea that loyalty is a fundamental military virtue. He challenges the idea that a soldier's primary loyalty should be to his or her fellow soldiers and explores the misplaced role that loyalty can play in military decision making. Olsthoorn offers an excellent analysis and reflection on what loyalty should mean to any member of the military. Conflicting loyalties can generate great confusion in a soldier's mind. Soldiers might feel that they are pulled in directions that are opposed to the values by which they are supposed to abide. They might also understand that loyalty to those values might be detrimental to their fellow soldiers. Olsthoorn argues that loyalty should be due only to a moral principle, not to individuals. Individual soldiers, when facing a conflict of loyalties, should therefore put loyalty to the moral principles underlying military service ahead of loyalty to their fellow soldiers, even if this means making decisions that might harm their fellow soldiers.

Bibliography

Clarke, W. and Herbst, J. (1996), "Somalia and the Future of Humanitarian Intervention," *Foreign Affairs*, 75:2, 70–85.

Cook, M. and Syse, H. (2010), "What Should We Mean by 'Military Ethics'?," *Journal of Military Ethics*, 9:2, 119–22.

Gourevitch, P. (1998), *We Wish to Inform You That Tomorrow We Will Be Killed With Our Families* (New York: Picador).

Honig, J.W. and Both, N. (1996), *Srebrenica: Record of a War Crime* (London: Penguin Books).

International Commission on Intervention and State Sovereignty (ICISS) (2001), *The Responsibility to Protect: Report of the International Commission on Intervention and State Sovereignty* (Ottawa, Canada: International Development Research Centre).

Killmister, S. (2008), "Remote Weaponry: The Ethical Implications," *Journal of Applied Philosophy*, 25:2, 121–33.

Litz, B.T., King, L.A., King, D.W., Orsillo, S.M., and Friedman, M.J. (1997), "Warriors as Peacekeepers: Features of the Somalia Experience and PTSD," *Journal of Consulting and Clinical Psychology*, 65:6, 1001–10.

Lucas, G. (2009), *Anthropologists in Arms* (Lanham, MD: Altamira Press).

Miller, L. and Moskos, C. (1995), "Humanitarians or Warriors?: Race, Gender, and Combat Status in Operations Restore Hope," *Armed Forces & Society* 21:4, 615–37.

Thakur, R. (1994), "From Peacekeeping to Peace Enforcement: The UN Operation in Somalia," *The Journal of Modern African Studies* 32:3, 387–410.

Tripodi, P. (1999), *The Colonial Legacy in Somalia* (Basingstoke: Macmillan).

Tripodi, P. (2006), "When Peacekeepers Fail, Thousands Are Going To Die. The ETO in Rwanda: A Story of Deception," *Small Wars and Insurgencies*, 17:2, 221–36.

Winslow, D. (1997), *The Canadian Airborne Regiment in Somalia: A Socio-Cultural Inquiry* (Ottawa: Canadian Government Publication Centre).

PART I
New Visions of Just War Theory

Chapter 1

"New Wars," Terrorism, and Just War Theory

Jessica Wolfendale

In recent years writers in philosophy, international relations, and politics have suggested that modern conflicts involve new and unique features that render traditional ways of thinking about the nature of war and the ethics of war inadequate or even dangerously redundant. For example, Mary Kaldor (1999) and Paul Gilbert (2003) have argued that modern conflicts no longer conform to the traditional model of interstate conflict motivated by concrete political aims. Instead, modern wars are characterized by low-intensity intrastate conflicts motivated by "identity politics" (Gilbert 2003),[1] involving non-state actors, high civilian-combatant casualties, the participation of mercenaries, and the use of tactics such as terrorism and human shields (Gilbert 2003; Kaldor 2007; Münkler 2005). Since 9/11 the term "new war" has also been used to describe the War on Terror and the current forms of international terrorism (Gilbert 2003, 3–6; Margolis 2004; Münkler 2005).

Other authors have argued that the shape of modern warfare means that we must rethink or even suspend the normative framework provided by just war theory. The increasing asymmetry between state military forces and weak state and non-state actors have led some to argue for rethinking just war theory (Rodin 2006), while others point to a "moral asymmetry" between forces that uphold or at least recognize the laws of wars, and those that do not see themselves as bound by any moral constraints in the conduct of war (Baker 2005). In addition, commentators such as Joseph Margolis (2004, 403–4), have argued that just war theory's strict limitations on the resort to military force are "effectively moribund" and even dangerously irrelevant when applied to modern international terrorism. Developments in military technology, such as the increasing use of unmanned autonomous weapons systems, have also led for calls for revisions in the areas of both jus ad bellum and jus in bello (Evans 2010; Killmister 2008; Sparrow 2007).

How should we analyse these claims? The aim of this chapter is to assess the claim that new forms of warfare, particularly modern terrorism, require revisions to just war theory. In Section 1, I summarize the claims of new wars theorists,

---

1   Gilbert (2003, 7) characterizes the war on terror as "an ideological struggle between the Islamists and US allies to win over ordinary Muslims to the one kind of identity or the other."

and argue that many of the features claimed to be unique to modern warfare are neither new nor substantially different from the characteristics of earlier conflicts. In Section 2, I explore two genuinely new aspects of modern warfare—the creation of a global war against terrorism and the extreme inequality in access to military power that has emerged with the rise of the military dominance of the United States. In Section 3, I discuss the arguments that just war theory cannot accommodate the ethical issues posed by modern conflicts. I outline the normative framework of just war theory and, using terrorism as a case study, consider when new features can and cannot be accommodated within just war theory. Finally, in Section 4, I touch on the implications of revising just war theory for our attitudes to and acceptance of moral constraints on war. I argue that adopting a new wars approach has problematic consequences for our attitudes toward those whom we fight, and to the laws and moral norms that underlie just war theory.

1. The New Wars Hypothesis

The new wars hypothesis was originally developed by Mary Kaldor (1999) to describe post-Cold War conflicts such as the Bosnian civil war. Since September 11 2001, the terminology of new wars has also been used to describe the fight against international terrorism. Paul Gilbert, for example, describes terrorism as a typical new war, and Joseph Margolis (2007, 402–3) characterizes 9/11 as an "act of war" which requires us to redefine the very concept of war. This characterization was adopted by the Bush administration—former President George Bush referred to the fight against terrorism as a "different kind of war that requires a different kind of approach and different type of mentality" (Jackson 2007, 356). Military writers in the US also adopted the language of new wars, arguing that we are entering a "fourth generation" of warfare, in which insurgencies and terrorism "threaten to undermine Western ways of waging war" (Roxborough 2006, 51).

    The term "new wars" has thus been applied to a variety of conflicts, from insurgencies to civil wars to terrorism. Despite the different applications of the term, however, proponents of the idea of new wars make a number of similar claims about modern conflicts.

    First, new wars are depicted as involving conflict between non-state and irregular forces, or between non-state and state forces. New wars are therefore typically asymmetrical, involving irregular forces that may not have the means to conduct conventional warfare, particularly if they are fighting a powerful state such as the United States (Gilbert 2003, 9). As Edward Newman (2004, 175) explains, according to new wars theorists: "the actors [in new wars] are insurgency groups, criminal gangs, Diaspora groups, ethnic parties, international aid organisations, and mercenaries, as well as regular armies." This explains why modern terrorism is seen as a typical instance of this phenomenon, involving as it does numerous sub-state actors and criminal gangs operating across state borders in conflict with state military forces.

Second, new wars are described as being motivated not by the desire for concrete political aims such as control over a government or territory, but by "identity politics" (Kaldor 2001, 6)—claims of political representation or control based on "national, clan, religious or linguistic groups" (Kaldor 1999, 6).[2] According to Gilbert (2003, 6–7), new wars are wars:

> ... between or on behalf of peoples, in the sense that the sides in the war are identified in terms of identities that are given, not just through membership of states or potential states, but through membership of groups whose representatives advance political claims ... in virtue of the kinds of people of which these groups consist.

For Gilbert, the war against terrorism is a paradigmatic case of a new war in which Americans are fighting "on behalf of those peoples who are taken to espouse the desired values [freedom and democracy]" and Islamists are fighting to protect Muslims from the influence of non-Muslims (2004, 7), an interpretation of modern terrorism that is shared by other authors (Huntington 1998). Fourth Generation warfare theorists also argue that new wars are conflicts between identities and values rather than conflicts over land or resources—conflicts between groups motivated by hatred and a desire to wipe out identities seen as hostile to their own. According to Ian Roxborough, Fourth Generation theorists believe that future conflicts "will be primordial in nature. They will pit one civilisation against another, and Western civilisation against barbarism" (Roxborough 2006, 51; see also Barnett 2003). It is frequently argued that this aspect of new wars (as well as the involvement of criminal gangs) makes it more likely that conflicts will continue, since criminal gangs and terrorist groups often have a financial motive for continuing profitable conflicts, and groups motivated by "identity politics" lack a clear political objective the achievement of which would mark the end of a conflict (Newman 2004).

Last, but not least, new wars supposedly involve a high civilian-combatant death ratio and the frequent use of tactics that are forbidden under the laws of armed conflict—"extreme and conspicuous atrocity," in Kaldor's words (2001, 99). New wars, it is argued, are usually fought by irregular forces (guerrillas, insurgency groups, terrorists) who target civilian populations directly and indirectly through such tactics as ethnic cleansing, terrorist attacks, hostage taking, systematic rape, hiding among civilian populations, and the use of human shields. Thus, the typical new war conflict is claimed to involve a significant disregard for if not complete dismissal of the traditional laws of armed conflict (Gilbert 2003, 9)—and thus a moral asymmetry between the two sides. Indeed, some authors argue that the weak states or non-state groups involved in new wars *deliberately* violate the norms of armed conflict in order to provoke their enemy into violating those norms

---

2   See Newman (2004), 174–5, for a helpful summary of the claims of new wars theorists.

themselves, thus undermining their enemy's legitimacy and moral superiority in the eyes of their population and the international community (Skerker 2004, 28–30).

In summary, new wars theorists depict modern conflicts—particularly the war against terrorism—as more dangerous, more atrocious, and longer-lasting than conflicts prior to the end of the Cold War. The rise of failed states, terrorist groups, and other sub-state actors has, it is argued, created a situation where failed states and weak non-state groups have no motivation to uphold the laws of armed conflict when they pursue their aims through conflict. As Margolis claims:

> We are moving into a world in which ordinarily disadvantaged, otherwise weak, poor, impoverished, helpless populations may have found a way to transform themselves into a force that might conceivably topple or severely wound even a superpower. (2004, 403–4)

Is this picture of modern warfare accurate?

1.2 Are New Wars Really New?

There are two implicit and sometimes explicit assumptions behind many of the claims made by new wars theorists. Firstly, it is assumed that both sides in conflicts in the first half of the 20th century upheld jus in bello constraints to a much higher degree than is currently the case. Secondly, it is assumed that states—particularly the US—uphold (or at least recognize the importance of) the moral constraints on war while the non-state groups they fight against do not. As I shall demonstrate, both this assumptions are mistaken.

Are wars more barbaric now than in the past? The previously quoted descriptions of new wars as "primordial," pitting "western civilization against barbarism," and characterized by "extreme and conspicuous atrocity" create a vivid image of modern war as a battle between civilized forces doing their best to uphold the laws of war, and uncivilized, barbaric groups who deliberately flout morality in pursuit of their own (irrational) agenda.

Striking though this picture of modern warfare might be, a brief survey of military history reveals that this and other empirical claims of new wars theorists are misleading, if not simply false. Firstly, contrary to the claims of new wars theorists, the number of intrastate conflicts has *declined* since the 1990s, and in fact *inter*state conflicts have risen since 1997 (Newman 2004, 180). Secondly, an analysis of civilian-combatant death ratios does not support the contention that modern wars involve a higher rate of civilian deaths and deliberate attacks on civilians than earlier wars. Once we take into account the deliberate attacks on civilian targets in World War II, the Second Philippines War of 1899–1902, and the colonial wars in Africa and Asia during the 1950s and 1960s, it becomes clear that *both* sides in many conflicts throughout the

20th century deliberately targeted civilians. These earlier wars also involved very high civilian-combatant death tolls (ten to on in the case of the Second Philippines War) (Henderson and Singer 2002, 165–90). There is therefore no evidence to support the belief that there has been a steady increase in civilian deaths and the deliberate targeting of civilians over the last 20 years.

Modern wars are also not significantly more atrocious than previous wars. World War II and the Vietnam War, to give just two examples, involved the deliberate use of genocide, terrorism, reprisal killings, rape, and systematic torture by *states* (including the US).[3] Deliberate violations of the laws of armed conflict by both state and non-state actors were frequent in many other conflicts throughout the 20th century (Melander, Oberg, and Hall 2006). As Edward Newman (2004, 181) succinctly puts it:

> ... there is no temporal, qualitative shift in the use of atrocity across the 20th century ... Warfare in the 20th century did not move from an ethos of chivalry among uniformed soldiers to one of barbarity among warlords and militia.

Indeed, many of the tactics that new wars theorists discuss in reference to the methods of non-state groups have been used with far more devastating effect by states throughout the last century.[4] For example, the death toll from acts of *state* terrorism, including the death toll from the World War II bombings of British, German and Japanese cities, and the death toll from the bombing of Cambodia during the Vietnam War, far outweighs the death toll of all acts of non-state terrorism combined, even when we include the attacks of September 11, 2001.[5] *Contra* the claims of many new wars theorists, then, widespread and deliberate violations of the laws of armed conflict are neither new features of conflict, nor the province of non-state actors—states have shown themselves equally willing to violate the laws of armed conflict. It is thus hard to sustain the belief that we are facing an unprecedented situation of moral asymmetry in today's wars (see Baker 2005), particularly since the US used torture and assassination in the war against terrorism.

---

3   The US conducted an illegal program of torture and assassination during the Vietnam War known as the Phoenix Program (McCoy 2006).

4   Following many commentators on terrorism, I define terrorism as the tactic of threatening or using violence deliberately against civilians (or non-combatants) in order to promote a political or ideological agenda. I do not have space to argue for this definition, suffice to say that this definition, like many others, allows for the possibility of both state and non-state terrorism. For more detailed discussions of the definition of terrorism, see Primoratz (2004) and Rodin (2004).

5   Primoratz and Rodin (amongst others) argue persuasively that philosophical definitions of terrorism should allow for the possibility of state terrorism.

2. What is New about "New Wars"?

While many of the features of modern wars that have been claimed to be new are not, as a matter of fact, new, I believe that there are two aspects of modern conflicts that are truly unprecedented.[6]

2.1 A Global War against Terrorism

Since the terrorist attacks of 9/11 many authors have argued that modern non-state terrorism presents a new and far more dangerous threat than ever before—a claim that, as noted above, does not bear up under scrutiny (Jackson 2005; Wolfendale 2007). However, while it is questionable that the threat posed by modern non-state terrorism is significantly greater than in previous decades, the characterization by the Bush administration of the fight against terrorism as a global "war on terror" was new. While several military dictatorships in Latin America used military force to combat what they characterized as internal terrorism, this was the first time a state had announced and prosecuted a worldwide war against terrorism that involved the invasion of other states and the assistance of numerous military allies.

Thus, describing the fight against terrorism as a "war" was not simply a rhetorical device, but referred literally to what David Luban called the "war model" approach to terrorism. As Luban argues (2002, 9–14), interpreting terrorism as an act of war rather than as a criminal act has important implications for the methods of counterterrorism used to fight it. In war, for instance, lethal force may be used against anyone who is part of the enemy forces, and unintentional proportional civilian deaths may be permissible.[7] So describing the fight against terrorism as a war was used to justify the use of war tactics and strategies—the deployment of military forces, the internment of suspected terrorist suspects without trial, and the acceptability of a certain level of risk to innocent civilians—that would never be permitted under traditional police or law enforcement models of fighting terrorism. As I shall explain in Section 3, describing the fight against terrorism as a war has radical implications for how the normative concepts of just war theory are understood.

---

6   The development of new autonomous weapons systems has also been claimed to be a new aspect of modern warfare. However, many of the ethical issues posed by these weapons, important as they are, are not significantly different from those raised by any long-range weapons systems. Unfortunately, space does not permit an in-depth discussion of these issues. See Sparrow and Evans in this volume for a discussion of new weapons systems.

7   Luban argues that the Bush administration adopted the permissions of the war model, and the permissions of a law approach (e.g. denial of the criminal's right to fight back) without the corresponding constraints.

2.2 Inequality of Access to Military Power

States, by virtue of their greater access to and control of military forces, have always been able to inflict far more damage than non-state groups when they resort to the use of military force. However, the 21st century has witnessed an unprecedented concentration of military power in the hands of a single country—the United States of America. In 2002 the US accounted for 39 percent of the world's total military expenditure, a figure that is certainly higher now. According to David Rodin (2006, 155), this figure is:

> … a higher proportion even than that of the Roman Empire at the height of its power … Since the demise of the USSR, the USA has no 'peer competitor', a power capable of engaging it on roughly equal terms in a conventional war.

This concentration of military power in the hands of a single state has changed the nature of modern war by making it next to impossible for weak states or non-state groups who believe they have a just cause against the US to engage the US on anything like an equal footing using conventional military means (Rodin 2006, 155–6). As Rodin points out, when we consider the rise of asymmetric wars in the late 20th century:

> … the disparity between the capabilities and tactics of adversaries that defines asymmetric war is constituted as much by the means and capabilities of the conventional power as it is by those of the non-conventional, disruptive actor. (2006, 155)

It would be naive to assume, therefore, that the apparent rise of non-state groups and weak states using non-conventional means of warfare is a phenomenon that is independent from the rise of the US military hegemony. For a weak state or non-state group to attempt to fight the US using conventional military tactics would probably be suicidal. Iraq's rapid defeat by the US in the first Gulf War is an example of what is likely to befall any state or non-state group that attempts to confront the US through conventional means. To understand the normative impact of modern conflict, therefore, we must consider how this asymmetry of power affects states that are fighting non-state groups *and* how it affects the rights and permissions of non-state groups when fighting states. We must be careful not to bias the discussion of the normative implications of modern warfare in favor of states by assuming that non-state groups could never have just cause, or that the moral issues raised by asymmetric conflicts rest primarily with the tactics and weapons that weak states or non-state groups might use. If the US is able to use lethal weapons that are risk-free for the military personnel controlling them (see Sparrow and Evans, this volume), for example, this raises the possibility that non-state groups facing such weapons could arguably resort to non-conventional tactics in response. Thus the fact that such groups may not be able to fight powerful states such as the US using conventional means of war has implications for the application of jus in bello and jus ad bellum to such conflicts.

In summary, the features of modern conflicts that pose genuinely new issues in light of just war theory are the definition of the fight against terrorism as a global war, and the extreme imbalance in military power that has emerged over the last 20 years. Do these aspects of modern conflicts require revision or even rejection of just war theory?

3. The Normative Implications of New Wars

The first step in analyzing new aspects of combat in light of just war theory is to clarify the basic normative framework that underlies the norms of jus in bello and jus ad bellum.

3.1 The Moral Framework of Just War Theory

Just war theory is a combination of legal and moral principles that regulate and limit when and how military force may be used. Many aspects of just war theory have been revised and refined over hundreds of years in relation to both the legal guidelines on the conduct of war and the moral principles underlying just war theory.[8] However the historical basis of just war theory does not provide a reason for thinking that new developments in warfare cannot be accommodated by the theory. While an in-depth analysis of the history of just war theory is beyond the scope of this chapter, a summary of the central moral principles of just war theory will provide a sufficient background against which to assess the normative implications of new forms of war.

Many of the fundamental principles underlying the criteria of jus ad bellum and jus in bello, as they are codified in international law, rest on moral principles that can be applied to many new developments in warfare. The laws of war present reasons for refraining from or performing certain actions that are independent from whether both sides in a conflict obey the relevant laws. For example, the jus in bello principle of discrimination is arguably best understood as an imperfect codification of the moral principal that killing innocent people is wrong.[9] Thus the reasons for obeying the principle of discrimination are

---

8　Larry May (2007) offers a detailed discussion of the origins of just war theory. See also Bruno Coppieters and Nick Fotion (2002).

9　It is imperfect because, as several philosophers have argued, the categories of combatant and non-combatant as defined in international law do not accurately reflect the category of innocent and non-innocent people, even when we understand "innocent" to mean non-contribution to the war effort rather than moral innocence. See Jeff McMahan (2004) and McPherson (2004).

independent from whether or not the other side also obeys the rule against killing civilians. As Jeremy Waldron (2010, 25) explains:

> ... the conventional rule that prohibits attacks on civilians prohibits something that—apart form the laws of war—would already be a grave moral offense. The default position, apart from any convention, is that intentionally killing or attacking any human being is prohibited as murder ... Absent the laws and customs of armed conflict, *all* killing in war would be murder.

The laws of war are therefore not merely conventional, unlike (for example) laws regulating driving. Unlike the laws that criminalize murder and rape, the laws that regulate driving do not codify pre-existing moral constraints into law. Instead, they are examples of *malum prohibita*—laws that constrain what was previously an area of complete freedom in order to coordinate people's actions in a safe and predictable way.[10] Given that we need laws to regulate driving, we have good reason to follow the law that requires us to drive on the right, and there are good reasons for such a law to be enforced. But if enough people began to drive on the left and the law was no longer enforced, there would cease to be good reasons for driving on the right—indeed it would become outright dangerous to do so if no one else was doing so (Waldron 2010, 17–18). Thus the laws regulating driving are purely conventional—the reasons for obeying those laws *do* depend, at least to some degree, on whether everyone else obeys them as well.

However, the prohibitions contained in just war theory are not conventional in this sense. As Waldron argues, the laws of war typically prohibit actions that are *malum in se*—wrong in themselves. It is true that the laws of war are necessarily simplified in order to regulate the resort to and conduct in war. Thus, they are rules that reflect a compromise between the need for clear law and the complex moral principles underlying those laws. But given that the laws of war typically prohibit actions that are *already* morally wrong, the reasons for obeying those rules are independent of whether the other side obeys them. The permission to kill enemy soldiers in war, for example, arises from an *exception* to the general prohibition against killing—an exception based on the belief that killing in self-defense and defense of others is permissible (Waldron 2010, 25). Thus, even if we are fighting an enemy who deliberately targets civilians, this does not make it morally permissible for us to do so as well. Civilians, by virtue of their status as non-threatening, remain illegitimate targets of lethal force.

It would be a mistake, therefore, to interpret the laws of war as a list of inconsistent and outdated rules that were developed in an attempt to serve the interests of both sides in a conflict and to minimize the destruction of war (Brandt 1972). Instead, it is more plausible to interpret the laws of war as a framework of rules based on fundamental moral principles such as the prohibition against killing

---

10   For an excellent discussion of the differences between *malum prohibita* and *malum in se*, see Luban (2005).

innocent people, and the moral permissibility of self-defense, against which we can assess new features of combat.[11]

Indeed, a number of revisions to just war theory have occurred over the last 50 years that provide a useful guideline for how to approach new forms of combat. For example, in the 1990s it became widely accepted that the traditional jus ad bellum criterion of just cause could be extended to include the resort to war for humanitarian reasons.[12] The inclusion of humanitarian intervention in the jus ad bellum criterion of just cause is consistent with the underlying moral principle that lethal aggression is only justified in cases of self-defence or defence of others. Humanitarian intervention is arguably a form of defence of others, and thus while humanitarian interventions are a relatively new form of war, the moral permissibility of this form of war can easily be accommodated within the basic normative structure of just war theory because it is consistent with the fundamental moral principles of the theory.

Given this basic summary of the normative framework of just war theory, we are now in a position to consider the new forms of warfare that I identified in Section 2.

A War Against Terrorism and Preventive War

One of the consequences of the characterization of the fight against terrorism as a war has been a call for the inclusion of preventive war in the jus ad bellum criteria—war to prevent possible, but not imminent, terrorist threats.[13] After 9/11, the Bush administration announced a war against terrorism and marshalled many of the basic normative ideas from just war theory to justify a preventive war against countries that were believe to harbour or support terrorism (Crawford 2003; Kaufmann 2007). As Brian Orend explains (2006, 78), in 2002 the Bush administration:

> … makes the case that—in light of modern terrorism and its objectives of a spectacular, hard-to-predict instances of mass civilian slaughter—a responsible government must reserve the right to strike first if it will most likely prevent such instances from occurring.

---

11   Larry May (2007) argues that the principle of humane treatment also underlies many of the laws of armed conflict, particularly those dealing with the treatment of prisoners and those banning the use of weapons that cause unnecessary suffering. This would be consistent with my claim that the laws of war codify pre-existing moral principles.

12   See Gerhard Øverland and David Lefkowitz in this volume for a discussion of the duty to provide humanitarian assistance. For a discussion of the justification of humanitarian intervention, see Coady (2002).

13   This is to be distinguished from preemptive war, which has been typically considered a permissible form of war in just war theory and is legal under current international law.

Orend goes on to argue that preventive strikes can meet the criteria of a just war, although he argues that the invasion of Iraq in 2003 failed to meet these criteria (2006, 78–83).

If we assess the argument for preventive war in light of the underlying moral principles of just war theory, it rapidly becomes apparent that preventive war is inconsistent with these principles. Preventive war would permit the use of military force against a state or a non-state group that has not yet committed any aggressive action, and thus any analogy between preventive war and a right to self-defence or defence of others fails. My right to self-defence does not entail a right to use lethal force against someone who does not pose an imminent threat to me but merely might, at some point in the future, harm me. Since the direct targets of a preventive war have not as yet committed any act of aggression, the deaths of innocent civilians that would inevitably be killed in a preventive war cannot be justified by reference to a state's right to national defence. The deaths of civilians in war has typically been considered permissible *only* as an unintended proportional side-effect of legitimate war aims, and so causing the deaths of civilians to prevent a merely possible threat cannot be justified within the basic normative framework of just war theory (Lackey 2006).

Secondly, characterizing the aim of preventing terrorism as a war aim is deeply problematic. Three of the traditional criteria of jus ad bellum are last resort, proportionality and probability of success.[14] Since war is so destructive, if it is to be justified it should be a last resort, it must be in pursuit of a just aim that is achievable through military means, and it should not cause more destruction than it is intended to prevent. War should then be fought in such a way as to minimize harm to those who are not part of the war effort. But one of the most serious problems with the concept of a war on terrorism is that it is very difficult to envision what would count as the end of such a war. Terrorism is not a state or a group who can be negotiated with, and so an end of the war of terrorism would have to be the end of the terrorist threat (Luban 2005). But this is an impossible goal. Terrorism is a tactic that is used by numerous groups (including states), and even if Al-Qaeda and other terrorist groups were destroyed, other terrorist groups would emerge to take their place. Thus waging a war against terrorism is inconsistent with two of the central moral principles of just war theory—the principle that war must be a response only to an actual or imminent threat, and that war must have some chance of success in achieving peace. Furthermore, a preventive war against terrorism fails the criteria of proportionality because it causes immense destruction

---

14   However, probability of success is not included in international law as a requirement for resort to war, perhaps because it is biased against weaker states (Orend 2005). This is an important point, but arguably the requirement of probability of success is connected to the requirement of proportionality and so cannot be entirely ignored. If a war has no chance of success (even if it is in a just cause), than that war will cause immense destruction and will also fail to prevent the evil it was intended to prevent, which seems to violate the principle of proportionality.

in order to prevent the mere possibility of a future terrorist attack (an attack that is very likely to be cause fewer deaths than those caused by the war itself[15])—in the service of an aim that is difficult, if not impossible, to achieve through military means. Thus an "open-ended" war on terrorism (Buchanan 2006, 29) cannot be accommodated with the normative framework of just war theory without serious violence to the underlying fundamental moral principles.

Asymmetric Warfare and Suspending Just War Theory

The ethics of asymmetric warfare are complex, and beyond the scope of this paper.[16] For the purposes of this chapter, I want to consider one aspect of the current extreme asymmetry between non-state groups and states: whether it is permissible for states to suspend jus in bello constraints when fighting an enemy who flagrantly violates those constraints.

Some have argued that the war against terrorism requires not merely the revision but the temporary abandonment of just war theory. Modern-day terrorism poses such an unprecedented threat to national and international security, the argument goes, that states cannot afford to uphold jus in bello constraints on, for example, the use of torture against terrorism suspects and targeted assassinations (Allhof 2003; Dershowitz 2002; Kaufmann 2007). As former White House legal counsel Alberto Gonzales argued in a 2002 memo to George Bush:

> ... the war against terrorism is a new kind of war ... The nature of the new war places a high premium on other factors, such as the ability to quickly obtain information from captured terrorists and their sponsors in order to avoid further atrocities against American civilians ... In my judgment, this new paradigm renders obsolete Geneva's [the Geneva Conventions] strict limitations on questioning of enemy prisoners and renders quaint some of its provisions. (Quoted in Danner 2005, 84)

In this view, since terrorists themselves flagrantly violate the norms of jus in bello, it would be foolhardy and even dangerous for a state to uphold the laws of war when fighting such an enemy. In Margolis' words (2004, 304); "in speaking of a war against terrorism, it no longer makes sense to say that a modern state can actually conduct a war along just war lines."

---

15   Conservative estimates place the number of civilians killed as a result of the war in Iraq at between 96,663 and 105,408, and the number of US military personnel killed was 4,326 as of July 2009 (NPR 2010). 2,752 were people killed in 9/11.

16   David Rodin (2006) argues that states have a duty to observe the jus in bello constraints to a higher degree of stringency than is normally required, when fighting a weaker opponent—a duty arising from the fact that states have significantly greater ability to cause harm to civilians and also greater ability to restrict the use of force (given their more accurate weapons systems).

It is not immediately obvious how to interpret these arguments. It is certainly not *impossible* for a state to uphold jus in bello constraints when fighting a terrorist organization. A state's decision to use military force is a deliberate choice. It is not forced upon them. When confronted with a terrorist threat, a state could choose from many different counterterrorism approaches (for example, they could use international law enforcement instead of war). Even if military force is used, a state has many choices about how, where, and when to wield that force. The state could choose to only use lethal force when they are certain that their targets are in fact terrorists. A state could decide not to engage in preventive strikes, given the risk of being mistaken about the existence of a terrorist threat and the risk to innocent civilians posed by such strikes. The state could decide not to target terrorist headquarters or terrorist leaders if they are embedded within civilian residences. The state could opt not to use tactics or weapons that increase the risk to civilians, such as high-altitude bombing (as was used by NATO forces in Kosovo), or weapons such as cluster bombs. To claim that state *cannot* fight terrorism without violating jus in bello norms is therefore disingenuous and fails to recognize the numerous military, police, and intelligence resources that states possess.

Perhaps, then, the most plausible interpretation of the argument that fighting terrorism requires suspending just war theory is not that it is impossible for states to fight terrorism within the constraints of just war theory, but rather that states won't be able to effectively *defeat* terrorist organizations if they fight within these constraints. Perhaps upholding the jus in bello would "tie the hands" of states to such an extent that they would be unable to defeat terrorism or would expose themselves to extreme and unacceptable risks while doing so. For example, if a state was to scrupulously uphold the distinction between combatant and non-combatant, they would be unable to effectively launch attacks at civilian areas in which the terrorist organization was based, or would have to resort to strategies that significantly increased the risk to their own soldiers, such as opting for ground attacks rather than aerial bombardment. Thus, as Michael Skerker argues (2004, 30), "Given a desire to strike at a legitimate military target, or to advance troops or armor, the agent [the stronger side] appears now to be faced with two choices: forgo discrimination to achieve these goals, or forgo the advance."

This argument is strongest if we make two assumptions. Firstly, we must assume that the stronger side has a just cause that warrants the use of military force in the first place, and secondly, we must assume the stronger side has no other options available that could achieve its military aims.

Both these assumptions are deeply problematic. Traditional just war theory separates the question of jus ad bellum from jus in bello, thereby allowing the possibility that a just war could be prosecuted with unjust means, and vice versa (Walzer 2000). This means that even if a state has a just cause in fighting terrorism, this does not make the tactics they use just, and nor does the existence of a just cause permit relaxation of the constraints of jus in bello. Just war theory is not primarily utilitarian, but deontological, and so the existence of a just cause does not render permissible just any tactic that happens to serve that cause (Waldron 2010). That violating the jus in bello might result in a faster end to a war does not in itself justify those violations, since the jus in bello constraints come into force precisely when there is a temptation to break them. These constraints are not intended to assist the rapid pursuit of a just aim. Instead, they reflect the fundamental protections that innocent people (and enemy soldiers) possess and that may not be simply sacrificed for the sake of efficiency. As I argued earlier, the constraints of jus in bello are based on moral principles that are independent from whether both sides in a conflict uphold them. Arguably, the only possible justification for violating such norms would be if doing so were absolutely necessary to prevent what Walzer (2000) called a "supreme emergency"—a threat to the very existence of a nation or community.[17] International non-state terrorism does not currently present such a threat (Wolfendale 2007).

Second, this argument assumes a false dilemma. States, as I argued earlier, have choices in regards to how they fight terrorism. Indeed, their access to military, police, and intelligence resources give them far more options than would be available to a non-state group that wished pursue a just cause against a state. If using military force against terrorist groups is difficult to do while upholding jus in bello constraints, then this provides a reason for a state to pursue other methods of fighting terrorism, such as international police operations.[18] Given the strength of the moral constraints against killing innocent people without just cause, adopting a strategy that permits the deliberate violation of jus in bello constraints in pursuit of a goal that is probably unattainable through military means cannot be justified.

4. The Consequences of New Wars Rhetoric

In this chapter I examined the arguments of new wars theorists in light of just war theory. I argued that modern conflicts are not distinctively new either in regard to the level of atrocity they involve or the use of tactics that violate the laws of armed conflict. Instead, I argued that modern warfare is characterized by the emergence of the concept of a war against terrorism, and by extreme inequality in access to military power.

Considering these aspects of modern warfare in light of just war theory was instructive in two respects. Firstly, once we understand the basic normative framework of just war theory, it becomes apparent that many of the suggested revisions to or suspension of just war theory cannot be justified. Revisions of

---

17   Some authors are sceptical that supreme emergencies do provide a justification for overriding jus in bello constraints (Coady 2004). However, even if one does accept the supreme emergency argument, the current threat of non-state terrorism does not meet the criteria of a supreme emergency.

18   See Coleman in this volume for a discussion of non-military means of fighting terrorism.

just war theory are only justified if they are consistent with the underlying moral principles of just war theory. Thus, I argued that characterizing the fight against terrorism as a war was inconsistent with the jus ad bellum criteria that war must be a last resort, must meet the requirement of proportionality, and must have some chance of success. Similarly, the argument that fighting terrorism requires the suspension of jus in bello prohibitions fails to recognize that states, by virtue of their greater access to military and police resources, have many options in responding to terrorism beyond using military force. Furthermore, the moral force of jus in bello prohibitions is not based on a requirement of reciprocity—the moral constraints of jus in bello hold regardless of whether both sides obey them, and so the fact that terrorist and other non-state groups flout these prohibitions does not generate any permission for states to do so as well.

Finally, I want to suggest that adopting the rhetoric of new wars encourages problematic attitudes towards the moral norms of just war theory. The nature and methods of war change over time, and one of the tasks of just war theory is to try to impose moral constraints on what is an extremely destructive enterprise—indeed, an enterprise that has become, if anything, far more destructive over the last one hundred years. Adopting more permissive norms in relation to the resort to military force and the conduct of war could increase the likelihood of resort to military force and reduce incentives to consider alternative methods of dealing with new forms of political violence (Buchanan 2006, 3; Kaldor, 2003). Therefore, if we believe that new threats, new forms of conflict, and asymmetries of power mean that just war theory is redundant or that the moral constraints in just war theory must be weakened, then we threaten the already tenuous commitment to moral restraint in war.

The rhetoric of new wars also encourages a view of the enemy as barbaric, irrational, and inhuman. This perception of the enemy can alter how state military forces understand their professional self-conception. Rather than considering the enemy as deserving of respect and restraint, military forces can come to see themselves as fighting an enemy who is ruthless, alien, inhumane, and desperate to kill them (Roxborough 2006). When combined with the belief that jus in bello prohibitions might need to be suspended when fighting such an enemy, this perception is likely to lead to violations of the laws of armed conflict and a significant disregard for the lives of enemy soldiers and the lives of civilians who are killed in war.[19]

In conclusion, when we consider the numerous challenges raised by modern conflicts, it is crucial not to lose sight of the fundamental moral principles upon which just war theory is based. Revisions to just war theory *must* be consistent with those fundamental principles, or we run the risk of undermining the normative

---

19   Arguably, the fact that the US military is not keeping track of civilian deaths caused by the wars in Iraq and Afghanistan is indicative of a failure to be deeply committed to the prevention of civilian deaths.

structure of the theory—a theory that is, at best, only occasionally observed in even the most civilized wars. As Jeremy Waldron (2010, 24–5) argues:

> Sustained violations [of the laws of war] … or the development of and implementation of doctrines that hold the laws of war in contempt stand a good chance of adding so much to the number of violations that the convention has to reckon with anyway, that they will contribute significantly to the failure of the entire enterprise.

Accepting just war theory means accepting that moral constraints on the resort to and conduct of war hold even when it seems more beneficial or efficient to break them. If we believe that the norms of just war theory may be sacrificed to efficiency or because they appear old-fashioned, then this shows we have not understood what it means to be committed to a moral principle.

Bibliography

Allhof, F. (2003), "Terrorism and Torture," *International Journal of Applied Philosophy* 17:1, 105–18.

Baker, D. (2005), "Asymmetrical Morality in Contemporary Warfare," *Theoria* 52:106, 128–40.

Barnett, R.W. (2003), *Asymmetric Warfare: Today's Challenge to the U.S. Military Power* (Washington DC: Brasseys Inc.)

Brandt, R.B. (1972), "Utilitarianism and the Rules of War," *Philosophy and Public Affairs* 1:2, 145–65.

Buchanan, A. (2006), "Institutionalizing the Just War," *Philosophy and Public Affairs* 34:1, 2–38.

Coady, C.A.J. (2004), "Terrorism, Morality, and Supreme Emergency," *Ethics* 114:4, 772–89.

Coady, C.A.J. (2002), *The Ethics of Armed Humanitarian Intervention* (Washington, DC: United States Institute of Peace).

Coppieters, B. and Fotion, N. (eds) (2002), *Moral Constraints on War: Principles and Cases* (Lanham, MD: Lexington Books).

Crawford, N.C. (2003), "Just War Theory and the U.S. Counterterror War," *Perspectives on Politics* 1:1, 5–25.

Danner, M. (2005), *Torture and Truth: America, Abu Ghraib, and the War on Terror* (London: Granta Books).

Gilbert, P. (2003), *New Terror, New Wars* (Edinburgh: Edinburgh University Press).

Henderson, E. and Singer, J. (2002), "'New Wars' and Rumors of 'New Wars'," *International Interactions: Empirical and Theoretical Research in International Relations* 28:2, 165–90.

Jackson, R. (2007), "Language, Policy, and the Construction of a Torture Culture in the War on Terrorism," *Review of International Studies*, 33:3, 353–71.

Jackson, R. (2005), *Writing the War on Terror: Language, Politics, and Counter-terrorism* (Manchester: Manchester University Press).

Kaldor, M. (2006), "The 'New War' in Iraq," *Theoria* 53:109, 1–27.

Kaldor, M. (1999) *New and Old Wars: Organized Violence in a Global Era* (Cambridge: Polity Press).

Kapitan, T. and Schulte, E. (2002), "The Rhetoric of 'Terrorism' and its Consequences," *Journal of Political and Military Sociology* 30:1, 171–96.

Kaufmann, W. (2007), "What's Wrong with Preventive War? The Moral and Legal Basis for the Preventive Use of Force," *Ethics and International Affairs* 19:3, 23–38.

Killmister, S. (2008), "Remote Weaponry: The Ethical Implications," *Journal of Applied Philosophy* 25:2, 121–33.

Luban, D. (2002), "The War on Terrorism and the End of Human Rights," *Philosophy and Public Policy Quarterly* 22:3, 9–14.

McCoy, A. (2006), *A Question of Torture: CIA Interrogation from the Cold War to the War on Terror* (New York: Henry Holt and Company).

McMahan, J. (2004), "The Ethics of Killing in War," *Ethics* 114:1, 693–732.

McPherson, L. (2004), "Innocence and Responsibility in War," *Canadian Journal of Philosophy* 34:4, 485–506.

Margolis, J. (2004), "Terrorism and the New Forms of War," *Metaphilosophy* 35:3, 402–13.

May, L. (2007), *War Crimes and Just War* (Cambridge: Cambridge University Press).

Melander, E., Oberg, M., and Hall, J. (2006), "The 'New Wars' Debate Revisited: An Empirical Evaluation of the Atrociousness of 'New Wars'," *Uppsala Peace Research Papers No. 9* (Uppsala, Sweden: Department of Peace and Conflict Research, Uppsala University).

Münkler, H. (2005), *The New Wars* (Cambridge: Polity Press).

National Public Radio (2010), "The Toll of War," available from http://www.npr.org/news/specials/tollofwar/tollofwarmain.html

Newman, E. (2004), "The 'New Wars' Debate: A Historical Perspective is Needed," *Security Dialogue* 35:2, 173–89.

Orend, B. (2006), *The Morality of War* (Peterborough, Canada: Broadview Press).

Orend, B. (2005), "War," *Stanford Encyclopaedia of Philosophy, Fall 2008 Edition*, Edward N. Zalta (ed.), available from http://plato.stanford.edu/entries/war/

Rodin, D. (2006), "The Ethics of Asymmetric War," in Richard Sorabji and David Rodin (eds), *The Ethics of War: Shared Problems in Different Traditions* (Aldershot: Ashgate), 153–68.

Rodin, D. (2004), "Terrorism Without Intention," *Ethics* 114:4, 752–71.

Roxborough, I. (2006), "The New American Warriors," *Theoria* 53:109, 49–78.

Skerker, M. (2004), "Just War Criteria and the New Face of War: Human Shields, Manufactured Martyrs, and Little Boys with Stones," *Journal of Military Ethics* 3:1, 27–39.

Waldron, J. (2010), "Civilians, Terrorism, and Deadly Serious Conventions," *New York University School of Law Public Law & Legal Theory Research Paper Series* Working Paper No. 09–09. Accessed May 4, 2010 from http://ssrn.com/abstract=1346360.

Walzer, M. (2000), *Just and Unjust Wars: A Moral Argument with Historical Illustrations*, 3rd edn (New York: Basic Books).

Wolfendale, J. (2007), "Terrorism, Security, and the Threat of Counterterrorism," *Studies in Conflict & Terrorism* 30:1, 75–92.

Chapter 2

Just War, Irregular War, and Terrorism

Stephen Coleman

Introduction

In recent times there has been considerable discussion about whether just war theory is relevant to the modern world, especially in regard to whether the theory offers any guidance (or at least any useful guidance) in dealing with the types of conflict prevalent in the 21st century. In this paper I wish to examine some of the arguments that have been put forward in this area, with a particular focus on the problems of relating just war theory to terrorism and counter-terrorism, and to insurgency and counter-insurgency operations.

Both those who claim that just war theory can be applied to asymmetrical wars, such as Alex Bellamy (2006) and Daniel Zupan (2004), and those who claim that it needs to be modified in order to deal with such cases, such as Nick Fotion (2007), (2006) use the usual six principles of jus ad bellum and two principles of jus in bello to assess the justness of such conflicts. While it is certainly common, at least in the popular press, for the actions of those deemed to be terrorists to simply be condemned, writers using some version of just war theory to assess the merits of asymmetrical conflicts do use just war principles to assess the justness of war from both sides.

Since it is the longest and most detailed exposition of the problems that just war theory faces when dealing with asymmetrical conflicts, I will begin by describing Fotion's argument that just war theory needs to be modified in order to cope with the problems raised by modern war. However, it is not my intention to simply present and then rebut the arguments of Fotion and like-minded theorists, since examining these arguments and the possible problems with them allows me to make more general claims about the various ways in which just war theory might be applied to the wars of the 21st century.

Irregular Just War Theory

Fotion claims that, while just war theory has ancient roots, it has evolved in more recent times, at least since the Peace of Westphalia in 1648, as a tool to deal with the ethical problems raised by state-versus-state conflicts. Indeed, even before Westphalia the ideas of just war theory could be applied in a very similar way, since even though the modern state did not exist, the entities that waged war on each other possessed many "state-like" characteristics. They were also, in all the senses relevant to just war theory, symmetrical entities, so the theory could be applied equally to both sides in the conflict. Fotion argues that the theory may be in need of some modification to deal with modern times, especially with regard to the understanding of a just cause for war, but that overall it still deals quite well with the ethical issues of state-versus-state conflicts. However, many modern conflicts are not of this sort and Fotion claims that the theory needs to be modified if it is to give us any clear sense of how a state ought to act when the threat of war is not being generated by another state, but rather by some non-state entity, such as a rebel army, insurgent group, or guerrilla organization.

In fact his claim is that not one but two theories of just war are required; the regular just war theory (which Fotion labels JWT-R) for dealing with the ethics of traditional wars between states, and a second irregular just war theory (labelled JWT-I) for dealing with the ethics of modern asymmetrical wars, which he defines as those wars fought between nations and non-nation groups. JWT-I marks a significant departure from traditional just war theory, in that various aspects of the theory apply asymmetrically to the opposing combatants.

In recent times it has often been claimed, particularly by those fighting such wars, that traditional just war theory, and the modern Laws of Armed Conflict that are essentially derived from it, seem to place requirements on the conduct of national armies that do not apply to the non-national groups that they are fighting in such wars. Indeed it has been common to claim that applying the laws of war to such conflicts leaves the nation to fight with one arm tied behind its back while placing no such restrictions on the non-national groups that oppose them. Yet Fotion's claim is that this is a mistaken view of the situation, and that using JWT-I to assess the justness of an asymmetric war will reveal that all the advantages do not fall on one side. In examining the six principles of jus ad bellum Fotion claims that while there are some criteria that must be met before a state can be considered to be acting justly in making war on a non-state entity, not all six criteria apply to such conflicts, unlike in JWT-R where a state must meet all six criteria in order for their actions to be considered just. In addition he claims that while there are some criteria that must be met before a non-state entity can be considered to be acting justly in making war on a state, they are in fact different criteria that must be fulfilled than those which apply to the actions of the state in such conflicts; in other words, in asymmetrical conflicts, the principles of jus ad bellum actually apply asymmetrically as well. Fotion claims that a similar situation applies to the jus in bello principles, in that these principles also apply asymmetrically in asymmetrical conflicts. Fotion's claims (2007) about the six individual jus ad bellum principles are as follows:

1. Right intentions; applies equally to both sides.
2. Proportionality; applies equally to both sides.
3. Legitimate authority; must be met by the state in that the legally designated authority within the state must be the one who authorizes the use of military

force against the non-state entity (i.e. authorizes the "war"). However, Fotion claims that this principle does not usually (and plausibly even does not ever) have to be met by the non-state entity. A rebel group may have a leader—Fotion gives the example of Fidel Castro leading the Cuban revolutionaries in the 1950s—however that leader has no legitimacy in the sense required by just war theory, thus he suggests that a non-state group does not have to meet the legitimate authority principle since they cannot satisfy it in principle (2007). It may be the case that the rebellion comes to have legitimacy; if it transforms from a loose association of rebel fighters with a leader (or group of leaders) to being an army engaged in war on behalf of a newly formed government as was the case in the American Revolution, and in such a case Fotion suggests that JWT-I would no longer apply and that the requirements of JWT-R would then have to be met in order for the war to be considered just.

4. Likelihood of success; must be met by the state, but again Fotion claims that the non-state group does not need to meet this principle, since it is almost inevitably the case that at the beginning of a struggle, when jus ad bellum is supposed to be applied, that the non-state group would be assessed as having virtually no chance of success, no matter how well-intentioned or honorable their motives, or how worthy their cause.

5. Just cause; must be met by the non-state entity, where the just cause for launching an asymmetric war will be essentially the same as for launching a traditional war; i.e. as a response to aggression. Fotion does note that unlike in the case of JWT-R, where the aggression usually involves the crossing of national borders, in an asymmetric war the aggression which provokes the war will most commonly be internal, and borders will not usually be an issue. On the other side of the coin however, Fotion argues that states actually have more latitude with regard to just cause in asymmetric wars than they do in regular ones. Fotion claims that preventative wars are not permitted under JWT-R;[1] knowing that a foreign state (Ardania) is training forces and making plans to attack its neighbour (Bedania) at some unspecified point in the future does not give Bedania just cause to launch a war against Ardania at the present time, but Fotion argues that acquiring similar information about the plans of a non-state group would give the state just cause to launch an attack under JWT-I.

The nation is permitted to attack a non-regular enemy that has not necessarily been clearly identified and may not even have been responsible for some

---

1 Pre-emptive attacks are those launched when the enemy's attack is imminent, preventative attacks are ones launched when the enemy is expected to attack at some point in the future; exactly where the line ought to be drawn between the two is not always clear. Michael Walzer (1977) also discusses this issue, and argues in favor of pre-emption and against prevention.

not-so-recent attacks on it. If by chance or good intelligence it locates these irregular forces, it is allowed to attack them. Because of the great variation among the kinds of non-regulars a nation faces, the permission to attack preventatively has to be granted on a case-by-case basis. In general, if the non-nation group has powerful area weapons, is in the process of collecting more weapons, is in the process of gaining new recruits to its cause and/or has plans for a future violent event, then it can be attacked. To repeat, it can be attacked even if it has not initiated any attacks as yet (2007, 117).

Fotion's argument is primarily based on the difficulties of defining what it means to be "at war" in an asymmetric conflict. As he notes, if Ardania and Bedania are already at war, there is nothing wrong with Ardania engaging in attacks against Bedania's military forces which are designed to prevent Bedania from launching an attack against Ardania; attacking planes while they are still on the ground is a good example of this. If a state of war already exists, then preventative attacks are seen as good tactics; as Fotion notes "there is nothing immoral about preventing your enemy's *next* attack; what is immoral is trying to prevent the *first* attack" (2007, 103).[2] The problem in irregular wars is that it is often difficult to know whether a state of war exists or not. A good example of was the conflict between the Tamils and the Sinhalese in Sri Lanka. The levels of violence in the country fluctuated dramatically over a long period of time, and thus if an attack occurred after a long period of relative calm, it was difficult to know whether to characterise this as a new war or a continuation of the old one. This is in marked contrast to the Arab–Israeli conflicts, where despite ongoing disputes it is quite clear that the 1973 war, for example, was a new war and not simply a resumption of the hostilities from 1967. Even if a non-state group has not yet launched any attacks, Fotion argues that the fact that it is arming itself in preparation for launching an attack can be taken as a sign that a state of war already exists.

6. Last Resort; must be met by the non-state entity who need to make an honest attempt to resolve the dispute by non-violent means. However,Fotion argues that states are in many cases severely restricted in the options open to them in dealing with non-state entities. When dealing with another state one can threaten, blockade, sanction, embargo etc, but few of these options are likely to be available to a state when dealing with a non-state entity and thus military action may be not the last resort, but the only resort. Fotion's argument for the permissibility of a state engaging in preventative strikes against a non-state entity is also applicable in discussions of last resort. If Ardania acquires intelligence that a particular domestic rebel group is planning on launching an attack, the available options are far fewer than in a similar case involving two states, particularly if the rebel group is in

---

2   Emphasis in the original.

hiding. When dealing with another state, one knows where the enemy is, how to contact its leaders, and so on. But when dealing with a rebel group or other non-state entity, things are not so straightforward; it is extremely difficult to engage in the sorts of non-military measures that might be taken against another state when one is dealing with a hidden rebel group, and thus military action may be the only realistic option.

Fotion also discusses the two principles of jus in bello and argues that they also apply asymmetrically under JWT-I;

1. Proportionality; applies equally to both sides.
2. Discrimination; Fotion argues that the nature of irregular war means that the discrimination principle cannot be upheld to a high standard by military forces, since they are likely to have difficulty in identifying their enemies who usually do not wear uniforms, are likely to be hidden amongst the civilian population, and often choose to fight in places where there are large numbers of innocent civilians in the vicinity.

Fotion does claim that the principle of discrimination must be met by the non-state entity if they wish to fight justly, since they do not face the same difficulties that the state's military forces do, in that enemy combatants and facilities are easily identifiable, and can be attacked at a time of the non-state entity's choosing. He notes that rebel and insurgent groups in fact regularly violate the principle of discrimination, but he argues that the claims used by such groups to support relatively indiscriminate attacks are not justified under just war theory, either in its traditional form or JWT-I. There are two arguments that he examines in this regard, the first being the claim that all those who benefit from the actions of the rebel group's enemies ought to count as legitimate targets, which he dismisses as radically narrowing the definition of "innocent" for no good reason. The second argument that Fotion examines and dismisses is the claim made by many insurgent and rebel groups that because they are weak they have no choice about the targets they attack; they are not capable of attacking military targets and thus the only option available to them is to attack non-combatants.[3] Fotion notes that attacking "soft" targets will certainly be easier than attacking military ones, but he does not agree that such attacks are necessary for many, if not most, non-state military entities.

---

3   Michael Walzer's principle of supreme emergency is occasionally invoked as a defence of this position, such as by James Sterba (2005) who uses the principle to defend Palestinian suicide bomb attacks.

Problems with Irregular Just War Theory

Since I think few people will disagree with Fotion's claims with regard to those just war principles that *do* apply to those engaged in an asymmetric war, I wish to focus this discussion solely on those principles which Fotion claims do not apply; legitimate authority and likelihood of success on the rebel/insurgent side, and just cause, last resort and discrimination on the national military side.

Legitimate Authority and Likelihood of Success

I think there is something deeply problematic in the idea that a non-state entity could be justified in going to war without any consideration at all of whether those starting a war have any authority to do so, or any need to calculate their chances of actually succeeding in their aims. If Fotion's argument here is taken to its extreme, then it would be ethically justifiable under JWT-I for a single individual to declare war against a state, assuming that this individual was able to demonstrate a just cause, right intention, proportionality and last resort (something which I for one do not think is actually as improbable as it might sound).[4] Fotion's rejection of both of these principles does seem to be rather hasty.

Consider the likelihood of success principle. Fotion's argument here is that it will almost always be the case that a rebel or insurgent group will, at least at the start of their revolution, have to assess their chances of success as realistically as possible, and will almost always conclude that their chances of success are very low, and thus according to Fotion, beginning their revolutionary activity would thus be acting unjustly according to traditional just war theory; "Rebellions would automatically be immoral even if just cause were gloriously on their side" (2007, 89). I think the problem with Fotion's claim here is the manner in which he assesses the likelihood of success of revolutionary groups, for even while he is examining cases of irregular war he seems to fall into the trap of examining them in terms of traditional war. If Ardania is contemplating going to war with Bedania, then just war theory requires Ardania's leaders to assess their likelihood of success, which in a situation like this will involve calculations of the number of troops, ships, planes, tanks etc that each side has, the terrain where battles are likely to take place, and so on. But assessing the chances of success of a revolutionary struggle is, I believe, quite a different calculation.

The case of the struggle for the rights of black people in South Africa, primarily led by the African National Congress (ANC) is probably a good example here.[5] When the ANC made the decision in 1961 to engage in military actions in support of their struggle for equal recognition, any objective observer would have had to

---

4   This issue is discussed by Uwe Steinhoff (2007). In particular he supports the liberal idea that a community can only have rights which its members transfer to it which would suggest that an individual could legitimately declare war on a state.

5   This case is discussed by Alex Bellamy (2006).

assess their likelihood of military success as being very small. The South African military was well-trained and well-equipped, and despite having the support of the majority of the population, the ANC had little prospect of ever being able to muster the military might to tackle them in open combat. Thus according to Fotion's assessment, their likelihood of success was very small, despite just cause being "gloriously on their side." But of course it was never the intention of the ANC to engage in open combat, and nor was this necessary in order for their revolutionary struggle to succeed. Violent action was only one aspect of ANC strategy, and other tactics, such as mass strikes and international lobbying, were also integral parts of the revolutionary conflict. An independent assessment of the likelihood of success in 1961 that ignores all of these non-military factors, especially an assessment which ignores the fact that just cause was quite clearly on the ANC's side—a fact highly relevant in terms of swaying international opinion against the government's apartheid policies—gives a highly unrealistic answer on the question of likelihood of success.

There seem to also be significant problems with Fotion's claim that non-state entities do not need to meet the requirement of legitimate authority in order for them to be engaged in a just war. In examining the revolution in Cuba in the 1950s Fotion raises the possibility that the legitimacy of the revolution might be derived from the people; if the people supported the revolution then it was legitimate. However he dismisses this answer as being too glib, noting that while Castro talked as if he was the people's chosen revolutionary leader (and thus the legitimate authority) he was only one of several leaders who made such a claim. Yet dismissing the principle of legitimacy in this way also seems too glib; many people might claim to have the support of the people but there might only be one person who actually has the support of the people. Or it might well be the case that the leaders of several different groups engaged in irregular war against the same enemy all have legitimacy granted to them by the different sections of the population that they represent; this is essentially the situation under JWT-R in cases where a coalition of states all engage in a war against a common foe. One obvious example of this sort of situation in irregular wars would be the range of different resistance groups that engaged in guerrilla and underground war against the German occupation of France in World War II; each one of these different resistance groups had the support of some identifiable section of the population, and it seems perfectly reasonable to suggest that each group had legitimacy as a result of having the backing of that section of the population.

The question of legitimate authority for non-state entities engaged in warfare has been addressed by various other writers, such as Alex Bellamy (2006). He suggests there are three tests that must be passed in order for a non-state actor to demonstrate legitimate authority to wage war. First, the non-state entity must be able to demonstrate that they enjoy high levels of political support within an identifiable community. Second, they need to be able to demonstrate that this community shares the political aspirations of the non-state entity, including the strategy of violence which is being adopted. Third, they need to be able to make and uphold agreements with other organizations, as well as control the members of their organization, especially with regard to ensuring that those members abide by the principles of jus in bello;[6] the last point ties directly into the willingness and ability to make agreements with other organization, since the rules of jus in bello are in themselves an agreement between organizations. It could be argued that these conditions are extremely difficult for a non-state entity to meet, especially given that legitimate authority is a jus ad bellum requirement which ought to be assessed before any war commences. However, there have certainly been non-state entities who could claim to have legitimate authority under these criteria, and to have had it for a considerable time before they engaged in warfare against the state, such as the ANC or even, plausibly, the Palestinian Liberation Organization (PLO), so such an objection does seem to be a little overstated.

Just Cause and Last Resort

Fotion has also made a number of suggestions about the way that JWT-I changes the situation for states in dealing with non-state entities. I want to start by examining his claims that a state has the right to engage in preventative strikes against non-state groups and that military action can be a legitimate first, rather than last, resort. I believe that both of these claims again are rather hasty and in the end are unsupportable. Fotion's arguments in these two areas are not always distinguished, with discussions about the just cause of preventative strikes shading into discussions about why military action can be a legitimate first resort for a state in dealing with a non-state entity; however I will do my best to keep discussion of the two issues separate in this paper.

It should be noted that Fotion is in fact not always clear about the issue of last resort; in some points of his discussion he suggests that last resort simply does not apply to states when dealing with non-state entities, however at other points in his discussion he suggests that other options, such as negotiations, do in fact have a part to play. He notes that while it may be impossible for a state to negotiate with some groups because it simply does not know that they exist, the grievances of the section of the population that the group represents are almost certainly well known, since it is rare for a rebel group to launch into a program of violence without making their dissatisfaction with the government known in some other way. Once such grievances have become public it is possible for the state to negotiate ways in which such grievances can be addressed, since the state will probably know of political factions that are sympathetic to this cause. However, as Fotion makes clear, the situation is rather different from the case in JWT-R, for there a state negotiates with another state in an attempt to prevent war, and if the negotiations fail they will go to war with that same state, whereas in JWT-I the state negotiates with one party (the political factions sympathetic to the grievances of

---

6   Bellamy bases his third condition on arguments originally suggested by Janna Thompson (2002).

the non-state group) and then if the negotiations fail the state will go to war against a different party (the non-state group that holds the grievances). In effect what he is claiming here is that military action may be a legitimate first resort in dealing with the non-state group itself, but that military action will not be a legitimate first resort in dealing with the grievances of that group, as the grievances will normally be known to the nation's government, and be the subject of possible negotiation, at a time when the revolutionary group may not even exist. However, I think that even this claim is rather overstated, but in order to explain why I will need to turn to the issue of just cause.

The claim that Fotion makes with regard to just cause is that although preventative strikes are not permitted under JWT-R, a state acts legitimately in undertaking preventative strikes against non-state groups under JWT-I. There are two main arguments which he presents to support this claim. The first argument revolves around the fact that in irregular wars one has difficulty in distinguishing between pre-emptive attacks, which are of course permitted under any form of just war theory, and preventative strikes which are not. The second argument takes the fact that a particular non-state entity is arming itself as evidence that a state of war already exists between the state and that non-state entity; if this is true then what might appear to be a preventative attack against the non-state entity by the national military forces would in fact not be an attack designed to prevent the first enemy attack of the war, but rather an attack designed to prevent the enemy's next attack, which would thus make this a permissible action of the state, assuming that the strike complied with the requirements of jus in bello. I would argue that the first of Fotion's arguments is simply false, and that the second is based on an unjustified claim about the differences between regular and irregular war.

Fotion's first argument for the claim that a state is justified in engaging in preventative strikes against non-state entities relies, as I have said, on the difficulty in distinguishing between pre-emptive attacks and preventative attacks. But it can also be difficult to distinguish between pre-emptive and preventative attacks in conventional wars, or to distinguish between pre-emptive and preventative attacks by non-state entities against nations or national armies. The Israeli attacks against the Egyptian Air Force at the start of the so-called Six Day War in 1967 are often cited as classic examples of pre-emptive strikes, yet others argue that these attacks were preventative rather than pre-emptive. Other cases, like the US-led invasion of Iraq in 2003, seem to be much easier to define. In 1940 Germany invaded Denmark and Norway; their claim was that this was necessary in order to protect the neutrality of those countries from a proposed invasion by British and French forces, and in fact Norwegian neutrality was violated by Britain before it was violated by Germany. Do these attacks then count as pre-emptive ones (albeit against a third party) or as preventative attacks? The mere fact that it is difficult to distinguish pre-emptive and preventative attacks in conventional wars is not seen as a reason to allow preventative attacks under JWT-R, so it is rather difficult to see why Fotion believes that this difficulty is reason to allow preventative attacks under JWT-I.

One plausible response in defence of Fotion's position would be to suggest that the difference in JWT-I is that it is actually impossible to distinguish pre-emptive attacks from preventative ones, as opposed to merely being difficult (as is the case in JWT-R). However, this again seems to be overstating the case. Suppose our theoretical nation of Ardania is aware that several armed insurgency groups are forming in response to allegedly racist policies instituted by the Ardanian government, and then the state is actually attacked, with a truck filled with explosives being driven through the security gates of the main government offices and then detonated. The Ardanian intelligence services are not certain which of the various insurgency groups in actually responsible for the attack, but this does not seem to matter to Fotion, for according to him the Ardanian military would be justified in engaging in military strikes against all of the known armed insurgent groups, and in the circumstances just described he seems to claim that if the Ardanian military were to engage in military attacks against all these insurgent groups they could not know whether they were in fact engaging in preventative war or not. However, suppose that the Ardanian intelligence services had particularly good information about one of these insurgent groups, the People's Front of Ardania (PFA) and knew that not only was the PFA not responsible for the attack, but that the leaders of the PFA denounced the attack in front of their own supporters, citing the large numbers of innocent civilians killed in the attack and insisting that while military struggle against the Ardanian government was justified, that only legitimate military targets ought to be attacked, and that in any case attacks should not be carried out until it was certain that the disputes could not be resolved through political means. In such a case there does not seem to be any difficulty in determining that any attack by the Ardanian military against the PFA would be a preventative attack, rather than a pre-emptive one. While the case that I have given here is deliberately structured to make my point, similar cases are in fact not uncommon when states are dealing with a range of different insurgent groups, in that while the state's intelligence services may not know which group actually carried out a particular attack, they often know which groups did *not* carry out the attack. From this it follows that it is not impossible to distinguish between preventative and pre-emptive attacks in JWT-I, at least in some cases. If it is merely difficult, and not impossible, to distinguish between pre-emptive and preventative attacks in JWT-I, then this seems to be very similar to the case under JWT-R, and not sufficient reason to allow states to engage in preventative military strikes against non-state entities.

Fotion's second argument takes the fact that a non-state group is arming itself as evidence that a state of war already exists between the non-state group and the state with which they are in dispute.

> The rebels are not planning military action just in case certain things happen. Rather their plans represent the first step in a process that, eventually, will lead to violence. So they are in effect at war already with the nation they are opposing.

> Thus any military response to the rebels is preventative, but preventative of an
> attack that is in the process of being implemented. (2006, 59)

A claim like this seems to significantly overstate the differences between regular and irregular wars. Suppose Ardania and Bedania have been engaged in a long running dispute over the sovereignty of a particular island that lies in the strait between them. If Bedania was to start acquiring new and sophisticated weapons, and if Bedania's Generals started to discuss the best way to attack the units of the Ardanian military on the island in order to seize control of it, this would not give Ardania the right, under regular just war theory, to engage in an attack on Bedania, nor would this in itself constitute a declaration of war by Bedania against Ardania; if the mere possession of sophisticated weapons plus the existence of plans to invade another country constituted a state of war then the United States would probably have to be considered to be at war with most of the countries on the planet.

In the case quoted by Fotion above, I do not see why it should be assumed that because the rebels are planning military action that this means a state of war already exists between those rebels and the state they are opposing, since the rebels may have no wish for war, but rather may simply recognize that if other solutions fail that they will need to be prepared for violent action (even if only in strict self-defence). It may also be the case that the rebel leaders deplore violence, and hope never to have to use it, but recognize that weapons will be much easier for the rebels to obtain now, while there is still hope for a peaceful political resolution of the dispute, than will be the case if such political negotiations fail. Fotion's claim that the fact that a rebel group is acquiring weapons and making military plans means that violence is absolutely inevitable, and thus that a state of war already exists between the state and the non-state group, seems to massively overstate the differences between regular and irregular war and also to unjustifiably ascribe the worst of motives to the leaders of such rebel groups.

Discrimination

The last of Fotion's claims that I wish to examine is his suggestions with regard to the jus in bello principle of discrimination. Fotion argues quite extensively that this principle does apply to non-state entities engaged in irregular warfare against states; however he also claims that it is impossible for national forces to uphold the principle of discrimination to a high standard when engaged in irregular wars. The interesting part about this claim is that he really does not argue for it at all, but rather seems to think that everyone will take it for granted that this claim is true; in fact his discussion of this particular issue lasts exactly four sentences.

> However, in a struggle falling under JWT-I, national military forces face certain
> obvious difficulties: they often cannot identify their enemies; the rebels hide
> in rural and/or urban areas and do not wear uniforms; they look like innocent

civilians; they often choose to fight in places where real innocents are found in large numbers. These difficulties make it impossible for the national forces to honour the discrimination principle to a high standard. This does not mean that they can attack innocents intentionally, but it does mean that the amount of collateral damage allowed by the Rules of Engagement will be greater— inevitably so. The choice is either to allow more such damage or demand that nations simply not resist rebel groups and, in effect, let these groups win because they have placed themselves in the midst of the general population. (2007, 122)

Clearly the sorts of irregular conflicts that Fotion has in mind here are ones like the conflict between the Israeli Defence Forces and Palestinian insurgent groups in the occupied territories, or the conflict between US troops and insurgent groups in Iraq and Afghanistan. Obviously such conflicts are extremely significant in the modern world, but there are many other types of irregular conflicts where the types of conditions that Fotion is gesturing to here either do not apply at all, or only apply in a limited way. Given this, I do think it is dangerous to generalize in the way that Fotion has done here.

Of particular concern is the implied claim that the deaths of innocents as collateral damage in attacks by the national forces are not the fault of those national forces themselves, but rather are the fault of the rebels who chose to fight in a location that placed those innocents in harm's way. There are several reasons why I believe this particular claim is dangerous. If the national military feels that they do not hold any moral responsibility for collateral damage incurred in clashes with the rebel group(s) then they are less likely to take steps to limit that collateral damage.[7] It is also the case that the location of clashes is not always chosen by the rebel group, particularly in situations where the national military engages in preventative strikes, as authorized under Fotion's JWT-I; in these situations if the clash takes places in a heavily populated area it is because the national military forces, and not the rebel group, have chosen to bring the battle there. Finally, and of particular importance in cases where battles occur away from the home of the national military forces (e.g. US military forces in Iraq), Fotion's claim encourages those national military forces to see civilians in the vicinity of the battle as mere obstacles in the way of fulfilling the military objective, rather than as innocents who have a right to be protected from the conflict.[8]

---

7   David Rodin (2006) has argued that military forces equipped with sophisticated modern weaponry are actually more liable for collateral damage incurred in operations, rather than less liable as Fotion claims.

8   It is fairly obvious that the country of operation tends to make an enormous difference to what counts as acceptable collateral damage—World War II aerial bombing attacks on military targets that would be likely to cause extensive collateral damage went ahead against sites inside Germany, but generally were not considered against similar sites in occupied France. In a modern situation, it would seem highly unlikely that US military decision-makers would think that the sorts of collateral damage caused by a missile strike

Defining Terrorism

Given its importance in the modern context, I wish to finish this paper with a brief examination of whether the principles of regular just war theory can be applied to national forces engaged in a war against terrorism, despite the arguments to the contrary provided by writers like Fotion. One particular difficulty in dealing with the question of whether just war theory is relevant to discussions of terrorism and asymmetric war is in defining what these terms mean; in fact arguments about what actually counts as terrorism or terrorist activity are extremely common in discussions of the relevance of just war theory in the modern world.

The special issue of the *Journal of Military Ethics* focussed on the Israeli response to terrorism offers a very focussed and specific definition of an individual act of terror, which I will briefly discuss since the writers, like Fotion, are of the opinion that traditional just war theory cannot be applied to military actions against terrorism:

> An act, carried out by individuals or organizations, not on behalf of any state, for the purpose of killing or otherwise injuring persons, insofar as they are members of a particular population, in order to instil fear among the members of that population ('terrorize' them), so as to cause them to change the nature of the related regime or of the related government or of policies implemented by related institutions, whether for political or ideological (including religious) reasons. (Kasher and Yadlin 2005)

The authors clarify two specific points in this definition. In noting that their definition is restricted to individuals or organizations not acting on behalf of a state, they suggest that state sponsored acts of terror properly fall under the auspices of unmodified just war theory and since this type of conflict is not the focus of their paper they deliberately exclude state-sponsored terrorism from their definition.[9] In addition they are quite clear about the fact that according to their definition the targets of acts of terror are persons in general, and not necessarily non-combatants; thus it is possible under this definition for all the victims of an act of terror to be combatants, e.g. members of a front-line army platoon, and even combatants engaged in direct military activities, such as a combat patrol.

---

against a high value target in Iraq would be acceptable if the target were located in Atlanta, Georgia. Yet there does not seem to be any legitimate ethical reason for drawing such distinctions.

9   The authors argue that a state can act in a manner that terrorizes a particular population, but that such actions can be analyzed under the existing legal and ethical framework of just war theory. Thus a state which engaged in such a practice would be guilty of a war crime. However they do not discuss how such practices ought to be understood if they occur outside the bounds of traditional war.

One questionable aspect of this definition, given the insistence that the victims of a terrorist act can be combatants, is the difficulty in drawing any sort of distinction between terrorism, which is almost always deemed to be morally wrong, and guerrilla warfare, which is usually seen to be acceptable. Many of the attacks carried out by resistance fighters in Europe during World War II would seem to fit this definition of terrorism, for these attacks were carried out by members of a non-state organization, directed against members of a particular population (i.e. members of the German military and those seen to be collaborating with them), with the clear political aim of disrupting German control and, at least with regard to actions against collaborators, instilling fear. The objectives of guerrilla war can also be the same as the objectives of terrorist activity; indeed it could be argued that the aim of Palestinian activities is to end foreign occupation of their territory, which is the same as the aim of resistance activities in Europe in World War II. The authors claim that there are significant differences between guerrilla warfare and terrorist activity; they suggest that guerrilla warfare is conducted primarily for the purpose of disrupting military activity, and that guerrilla warfare is directed against military targets and military infrastructure. While they do not make the claim explicitly, it can be presumed that they believe that terrorist acts are not conducted for the same purposes or at the same targets, thus they conclude that guerrilla acts are very similar to military acts (and thus fall under the scope of traditional just war theory) and that terrorist acts are very different from military acts. However these differences certainly do not seem to be captured by their definition of a terrorist act.

I believe a much better approach is taken by Alex Bellamy, who notes that the terms "terrorism" and "terrorist" are not merely descriptive, but have a negative moral connotation. Bellamy thus begins by assuming that terrorism is immoral and then attempts to determine what specific characteristic of terrorism makes it immoral, for if such a feature can be isolated then this would seem to be a defining feature of terrorism.[10] Bellamy (2006) lists four characteristics common to most definitions of terrorism, and notes that most authors will point to one (or more) of these characteristics when trying to explain why such an action is wrong. The four characteristics are: (1) terrorism is politically motivated violence; (2) terrorism is conducted by non-state actors; (3) terrorism intentionally targets non-combatants; and (4) terrorism achieves its aims by creating fear within societies. Bellamy concludes that it must be the fact that terrorism intentionally targets non-combatants which makes it immoral, since all the other listed factors can be a part of legitimate military activities.

Bellamy's definition is at odds with Kasher and Yadlin's definition in several important respects, not least because defining terrorism as the intentional

---

10   One possible problem with Bellamy's approach is the very fact that he assumes that terrorism is immoral, for this effectively rules out the possibility of arguing, via something like Walzer's supreme emergency condition, that terrorism might, on occasion, be morally justified.

targeting of non-combatants allows for the possibility that states might also engage in terrorism. Defining terrorism in this way does fit in fairly neatly with just war theory in its traditional forms, for under traditional theory terrorism is clearly wrong as it violates the principle of discrimination. It should be noted that Bellamy does claim that even under this definition it is possible for military personnel to be victims of terrorist acts. An example that Bellamy (2006) points to is the 1983 bombing of the US Marine barracks in Lebanon. The reason that Bellamy claims this was a unjust attack is not because it failed to uphold the principle of discrimination, but rather because the war against the US Marines failed the jus ad bellum test, which he claims makes all attacks within that war immoral.

Conducting a Just War Against Terrorism

Given the manner in which terrorism has been defined, it perhaps makes little sense to speak of a war on terrorism, for the definition which I am using, as supplied by Bellamy, can only by applied to a particular action. It does make sense, however, to speak of war against a non-state entity which espouses terrorism as a tactic, so that is what I shall briefly discuss.

Fotion argues that just war theory needs to be modified in order to cope with irregular wars, but I do not think that any of the suggested modifications are an improvement on existing just war theory. All of the principles of jus ad bellum can, with a little thought, be applied to both sides of such conflicts, as can the principles of jus in bello. With regard to Fotion's claim that national forces are justified in allowing higher levels of collateral damage when engaging in a war against a non-state entity, there does not seem to me to be any moral justification for such a claim, in the same way that there is no justification for the non-state entity claiming that the principle of discrimination does not apply to them. Overall, I cannot see any way in which any of Fotion's claimed asymmetries can be justified. Having said this, I do think that there is one significant asymmetry which he has not considered.

In all of Fotion's discussions of irregular conflicts, he focuses on military engagement, which is unsurprising given the focus on just war theory. However this focus does mean that he overlooks another important aspect of a state's attempts to deal with rebellions, insurgencies and terrorism, the actions of those involved in upholding the law of the land, namely police and members of the criminal justice system. It is here that the real asymmetry lies, for while just cause might be gloriously on the side of the revolution, the law almost inevitably is not. In virtually every jurisdiction it is an extremely serious criminal offence to simply possess what might be termed "weapons of moderate destruction" let alone weapons of mass destruction. If military operations of a preventative nature are going to be justified – and in fact I agree with Fotion that they are – then they are going to be justified as military aid to the state's police forces, who are entirely justified in acting to prevent breaches of the law.[11]

There are other reasons for thinking that the most appropriate response to terrorism is a legal one, rather than a military one. If you are fighting a war, then enemy combatants who are taken prisoner ought to be treated as prisoners of war and as such they are not held personally responsible for their actions. If, however, the situation is treated as a police action aimed at arresting those who have broken the law, then the situation is quite different, for those who are taken into custody can quite properly be charged with any offences that they have committed.

I think it is particularly important to stress the policing nature of the operation to military personnel who are engaging in fighting an insurgency, especially when terrorism is a tactic being utilized by the opposition, for it is essentially impossible for a state's military to win such a war by military force alone. The aim of police and military operations in such a case is to reduce violence to a level where a political solution becomes possible. As Clausewitz noted, wars are not won by the side that inflicts the most damage; wars are won when one side decides it is not worth fighting anymore, and nowhere is this more obvious than in counter-insurgency operations. All the principles of just war theory can, and should, be applied to both sides of both regular and irregular wars; despite Fotion's claims there are in fact no moral asymmetries in asymmetric wars.

Bibliography

Bellamy, A. (2006), *Just Wars: From Cicero to Iraq* (Cambridge: Polity Press).

Fotion, N. (2007), *War and Ethics: A New Just War Theory* (London: Continuum).

Fotion, N. (2006), "Two Theories of Just War," *Philosophia* 34:1, 53–64.

Kasher, A. and Yadlin, A. (2005), "Military Ethics of Fighting Terror: An Israeli Perspective," *Journal of Military Ethics* 4:1, 3–32.

Rodin, D. (2006), "The Ethics of Asymmetric War," in Richard Sorabji and David Rodin (eds), *The Ethics of War: Shared Problems in Different Traditions* (Aldershot: Ashgate), 153–68.

---

11   In making the claim that preventative actions by police (and by the military in support of police) are the appropriate way to deal with terrorism, I am not making the claim that ALL actions undertaken by police against rebels, insurgents and terrorists fall into this category. It is easy to imagine a whole range of actions that might be taken by police against rebellions and insurgencies that, while legal, would not in any way be morally appropriate—many actions taken against the ANC by South African police and security forces during apartheid would appear to fall under this banner. Rather my point here is that the ONLY actions that could possibly be appropriate are actions taken by police (and those, including the military, who work in support of them) in order to uphold the law of the land. My thanks to Rob Sparrow for pointing out the need for clarification on this point.

Steinhoff, U. (2007), *On the Ethics of War and Terrorism* (Oxford: Oxford University Press).

Sterba, J. (2005), *The Triumph of Practice Over Theory in Ethics* (Oxford: Oxford University Press).

Thompson, J. (2002), "Terrorism and the Right to Wage War," in Tony Coady and Michael O'Keefe (eds), *Terrorism and Justice: Moral Argument in a Threatened World* (Melbourne: Melbourne University Press), 87–96.

Walzer, M. (1977), *Just and Unjust Wars* (New York: Basic Books).

Zupan, D. (2004), *War, Morality and Autonomy: An Investigation in Just War Theory* (Aldershot: Ashgate).

Chapter 3
The Jus Post Bellum

C.A.J. (Tony) Coady

Recent events in Iraq and Afghanistan have created renewed interest in those aspects of just war thinking that relate to the ending of war, what is often called the jus post bellum, on analogy with the jus ad bellum and the jus in bello. (For convenience, I will usually abbreviate these to JAB, JIB and JPB.) These aspects are worth treating, though dividing them off into what appears to be a quite separate category has its own problems, as indeed does the total separation of jus ad bellum and jus in bello. In what follows I shall be conscious of the connections as much as the contrasts between these (sometimes) useful terminological divisions.

The idea of "post bellum" is somewhat more ambiguous than it might seem, as the wars in Iraq and Afghanistan have demonstrated. When we turn our thoughts from JAB or JIB to jus post bellum we think of the ending of a war as a relatively discrete event or close-knit set of events, and some wars satisfy this characterisation. But others do not. Where some wars finish with surrender or a negotiated compromise that involves a definite date for the cessation of hostilities others take a different path. World War I finished with an armistice and a surrender on a definite date, and World War II followed a similar route. In Vietnam, a peace accord was negotiated in 1973 essentially between the Americans and the North Vietnamese, with the South Vietnamese government forced into signing by their American controllers. This allowed the Western forces to vacate the field without acknowledging the reality of defeat, leaving their local surrogates to deal with the breakdown of the treaty without the financial support they had been promised by the US and hence to face ultimate surrender. With Iraq, George Bush declared victory in 2003, so are we to take it that all the subsequent events in Iraq were not war but post-war reconstruction? This is, to say the least, a dubious description. In fact, in terms of violent fatalities, injuries and violence-related miseries, the period since "mission accomplished" has been far more damaging in many respects, if less spectacular, than the brief period of "the war."

This suggests one dimension in which the idea of an aftermath of war must be viewed with some flexibility; another dimension concerns the fact that thinking about the just ways the winners should proceed after the war raises questions about how they should seek to bring about the cessation of the war while it still continues. The post bellum arrangements must be governed by the ideal of peace but so must the ways of concluding the conflict, and the two phases are clearly related. Of course, the tradition of the just war has always held (if sometimes a little too quietly) that the ideal of peace is integral to the waging of war and has a role to play in the JAB and JIB so it is not surprising that it would have a prominent part to play in the JPB. I should stress at this point that my talking of two phases here indicates a decision to consider both the procedures for bringing hostilities to an end and the procedures for handling the strictly post-war activities of occupation, reconstruction and so on as falling within the one division or category of moral concern. I think there are sufficient related issues in the two phases to make this a plausible conceptual strategy, but at least one other author, Darrel Moellendorf, has recently divided them and, in a very interesting article, speaks of the *jus ex bello* (JEB) for the first phase concerned with ending the war and jus post bellum for the second phase (Moellendorf, 2008). I have sufficient sympathy for the illustrious Occam to resist multiplying *jus* beyond my necessities, so I will resist this terminological suggestion, though I will later address some of the issues that Moellendorf thinks justify it. There is however no great harm in proceeding as Moellendorf does.

Relations between JAB, JIB and JPB

There will in fact be many wars in which considerations of JAB, JIB and JPB are all relevant to what is going on, and what should be going on, in concluding hostilities and in establishing a new order between the warring parties. Elsewhere, I have argued against a strong thesis of the logical independence of the JAB and the JIB, here I would want to reject any similar claim that the JPB is independent of the JAB and JIB. This is clear enough for the idea of temporal independence since, as Iraq demonstrates, the considerations of the JPB can condition what is happening while a war is continuing in some form, and need not apply only after a war is concluded. Efforts at rebuilding civil and political society can be made (if with a whiff of paradox) even as hostilities of insurgent warfare or concerted terrorism continue in ways that can plausibly be said to extend the original war. But more generally this is only a dramatic instance of the logical connections between JAB, JIB and JPB. A war that is fought with a deliberate policy of violating the JIB cannot plausibly be held to satisfy the JAB (even if it has a just cause) since the determined violations at least invalidate the JAB condition of right intention. If this is how it is conceived in the first place then it should not be embarked upon, and if such systematic violations develop in the course of the war, they change its character and that reflects on its justificatory base in the JAB. As I (and others, notably Jeff McMahan), have argued there is also a problem concerning the permissions of the JIB with respect to the moral licence to kill enemy soldiers when your cause fails the conditions of the JAB. But these are issues for another occasion, though, as we shall see, they have some parallels with issues in the JPB. The present point is simply that a war that is fought without a considered view to bringing about a legitimate peace has a morally defective rationale that taints its legitimate beginnings and its ongoing processes. In passing, I might note that Moellendorf is anxious to establish the independence of his JEB and JPB from the

JAB, but our differences may be no more than terminological since I agree that the circumstances involved in facing the problems of ending or continuing a war and of establishing peace after a war may evoke moral considerations that are, in certain respects, different from those evoked in the circumstances of beginning a war. Nonetheless, there are ways in which the principles and considerations of the JAB and the JIB are pertinent to the JPB (in my sense), and this is something that Moellendorf agrees with, so his "independence" is consistent with the connections I insist upon.

Afghanistan is an interesting case for these connections, especially those of the JAB with the JPB. There is room for argument about the legitimacy of the US attack on Afghanistan in the first place, though many critics of the Iraq war nonetheless were, and remain, supporters of the Afghanistan intervention. So let us suppose that the JAB was satisfied and it was morally permissible to bomb the al Qaeda camps. The question then arises about the US extension of the conflict by joining the civil war against the Taliban government that was being waged by the Mujahadin Northern Alliance. The justifications offered for this were somewhat cloudy, but included the ideas that (a) the Taliban government was supporting and harbouring the al Qaeda forces so that the Taliban and its forces were legitimate targets for retaliation for 9/11, (b) the Taliban were running a brutal, repressive regime that deserved to be overthrown, and (c) the Taliban were actually complicit in the attacks of 9/11. Of these, (c) is not supported by evidence, unless it is merely a version of (a). As for (a), its validity depends upon disputed issues about the extent of the Taliban's involvement with al Qaeda, but if the Taliban was mostly tolerating al Qaeda's presence within its remote borders by not engaging in the difficult, dangerous task of trying to expel them (as seems possible) then it is hard to make the case for (a), even with respect to just cause, still less the other conditions of the JAB, such as proportionality, last resort and prospect of success. The case for (b) depends upon a humanitarian intervention argument that seems to me very difficult to establish, and that was not really made explicit at the time, though it figured somewhat in post hoc justifications (just as a parallel vindication was often produced regarding Iraq when the primary case for the war fell apart.) Difficult, because the oppression of women and other serious human rights violations by the Taliban government, abhorrent as they were, were not so extreme as to justify the upheavals that the expansion of foreign military involvement would require (and have indeed required in the event). In fact, the more plausible explanation (as opposed to justification) for the broadening of the offensive was the reluctance then of the US to commit ground troops to the attack upon al Qaeda strongholds, a reluctance that made a compact with the Northern Alliance to use their ground troops seem expedient. It was the momentum of this compact that basically drew the US and its Western allies into the project of overthrowing the Taliban. If this account of things is correct, then the issue of JPB is clouded by the dubious morality of the project itself, and moral questions to do with reconstruction, new political arrangements and so on become parallel to those involved in Iraq.

It should be stressed, nonetheless, that the question of how the perpetrators of an unjust war should end it and what they should do if they win such a war and face issues of occupation, reconstruction and so on is not immediately answered by the fact that their war was unjust. Even where they recognize that their enterprise is unjust they may have accumulated responsibilities and created complex, morally loaded situations for themselves and their opponents that require more nuanced responses than simply stopping or going home. This is a matter for discussion later.

The Scope and Perspective of the JPB

There are at least three characteristic areas of concern that need to be addressed under the heading of the JPB, all of them governed in one way or another by the just war concern with the ideal of peace. The second of them breaks up into several distinct sections or sub-areas. The areas are:

(a) How to end a war justly. Here, there are important questions to do with the role of negotiations, the preservation of enemy political structures to enable peace negotiations, the problems raised by policies of "unconditional surrender," and so on.

(b) How to deal with post-war conditions for the treatment of the defeated side by the victorious side. The sub-areas here are:

1. Rebuilding and reconstruction issues. How should the victors treat the defeated armed forces, the infrastructure of the defeated country, the government? What difference should the ideology of the victors make?

2. The question of punishment and reparations and the relation of this to truth and reconciliation processes.

3. The difficult roles of the military as occupiers, peacemakers and/ or peace-keepers. This is pertinent to the victors' role in post –war situations but also to peace-keeping operations where the peace-keeping forces haven't been making war, but there have been war-like situations on the ground prior to their arrival or intervention. Once more there may be lessons from truth and reconciliation institutions. But this is not a topic that I shall be able to address today.

(c) How to deal justly with one's own troops after the war. This is a very important, though somewhat neglected, issue. It is important for several reasons. One is the psychological and physical damage that surviving troops have suffered. Nations, or groups, that fight a just war owe a great deal to their soldiers who survive, and of course to their families and the families of those who do not. The issues are rather different for the post-war treatment of unjust warriors by their own country, and by other countries. This raises similar problems for the JPB to those created by what has been called the problem of asymmetry in the JIB (and discussed by Jeff McMahan, David Rodin, Henry Shue, myself and others in recent years).[1] This challenges easy acceptance of the moral permissibility for unjust warriors killing just warriors, a permissibility that is legally enshrined in international law and much modern philosophical writing on just war. Again, this topic of the right attitudes and appropriate treatment of one's own and enemy troops is one that I will not be able to address here.

Following one central strand in the just war tradition, I will initially develop my account of (a) and (b) 1 and (b) 2 in terms of what is required of the just side in a war with respect to the ending of the war, the establishment of peace in post-war circumstances, and the other issues mentioned above. This will also raise questions about what is permitted to such victors with respect to reconstructing the political arrangements of the conquered nation. There are a number of problems raised by the relevance of such an account of requirements and permissions to the question of what should apply to any victorious side, just or unjust, in such circumstances, and hence what are the best regulations or laws to promote for the phases of JPB.

A further consideration about one's perspective on the moral issues to do with war's end and aftermath is that, viewing the matter as I shall (at least initially) from the viewpoint of what is required and permitted for the just side, carries with it certain dangers, especially in the present political climate. These are dangers associated with moralism, an idea that I have explored elsewhere and which should be sharply distinguished from morality proper.[2] In connection with JPB, the stance of the just or putatively just warrior is particularly fraught with the prospect of moralistic distortion, especially when it comes to determining what should be allowed by way of reconstitution of the enemy polity. Consciousness of one's being justified in war combined with zeal for a particular political outlook or ideology can lead to illicit or imprudent imposition of reconstruction policies that not only work against peace but deny people a legitimate autonomy. As a convinced democrat, I have no doubt that democracy is a good thing, and only wish that it were taken more seriously by avowed democracies. As a convinced Christian, I have much the same view of Christianity. Chesterton's *bon mot* about Christianity that it "has not been tried and found wanting, but it has been found difficult and not tried" has its parallels for the case of democracy. Nonetheless, taking the opportunity afforded by a successfully waged just war in order to forcefully impose either democracy or Christianity suffers the defects of ideological imperialism.

Of course, it is unlikely these days, in spite of the religious impulses of some Western leaders, that the aftermath of a war waged by Western powers will produce religious imposition, but it is all too likely that it will bring some version

---

1 See, for instance, the essays in Rodin and Shue (2008).

2 For fuller treatment of the moralism concept see Coady (2008b), especially Chapters 1 and 2.

of imposed democracy. Considering the close connection between professions of democracy and agendas for economic advantage and control amongst those in power making such professions, it is extraordinary how naïve much philosophical writing on the privileges of righteous reconstruction appears. Brian Orend, for instance, who has probably written more than any other philosopher on JPB issues and is at least aware of some of the difficulties towards which I am gesturing writes in 2007 with apparently no sense of irony of the benefits of foreign investment that started to flow into Iraq in 2003! (Orend, 2007, 586) The commercial rip-offs and staggering corruption unleashed by the invasion and exploited by foreign security firms cum mercenaries, building contractors, and oil interests seem simply to have passed him by. Orend does not talk so much of democracy, but of imposing "a minimally just" and "rights-respecting government." When the victorious power itself has trouble meeting "rights respecting" criteria and has even more difficulty in recognising its own defects, the prospects for imposed "minimal justice" may be somewhat bleak.

Consider for example the recently much-publicized announcement by President Obama of a withdrawal from Iraq at the end of August 2010 "as promised and on schedule." According to the *Guardian*'s highly regarded correspondent, Seumas Milne, the withdrawal is nothing of the kind. Milne argues that what we now have is not an end to the war but a continuing occupation under another name. There will still be 50,000 US troops in 94 military bases which is a major reduction but still a huge number of troops. They will be "advising" and training the Iraqi army, "providing security" and carrying out what are called "counter-terrorism" missions. As Major General Stephen Lanza, the US military spokesman in Iraq told the *New York Times*: "in practical terms, nothing will change." At the same time the "privatizing" of the war with mercenaries, security contractors and the like goes on apace; there were in August 2010 about 100,000 private contractors working for and with the coalition forces of whom an estimated 11,000 are armed mercenaries. But as troops are removed these numbers are likely to escalate into what Jeremy Scahill, who helped expose the role of the notorious US security firm Blackwater, calls "the coming surge" in private contractors. Meanwhile, in spite of the way media attention has moved from Iraq to Afghanistan, civilians are still dying in significant numbers; in July 2010 alone there were 535 Iraqi civilians, the worst figure for two years. Moreover, a dozen 20-year contracts to run Iraq's biggest oil fields were handed out in 2009 to foreign companies, including the Anglo-American consortiums that exploited Iraqi oil under the British in the "good old days." [3] If this is what the provision of "minimal justice" looks like then the country is probably better off without it. Milne's picture and his interpretations may well be contestable, but the facts he cites suggest that, even at the level of principle, we may do better to be less utopian, less lofty, and less consumed by our

---

3   This paragraph summarizes part of the report in Milne (2010).

own righteousness, in prescriptions and principles for reconstructing conquered nations.[4] I will look more carefully at this issue later.

A point that should inform any treatment of the three areas mentioned above is that we must assume that a majority of the conquered people, or at any rate a significant proportion, are not primarily to blame for waging an unjust war. This outlook is already involved in the JIB principle of discrimination in targeting which exempts from attack those who have had no significant agency role in waging the unjust war, and are thus non-combatants. It shows one way in which the different categories of just war thinking are interrelated. Moreover, any post-war settlement or reconstruction must take account of the fact that very many of the enemy population who did play a significant part in waging the war will have been non-culpably ignorant of the injustice of their cause. Others may have been coerced into their participation, either as soldiers or workers in directly war-related industries. And most of those who were not ignorant or coerced may be indistinguishable to outsiders from those who were. These facts are pertinent to many policies generated by the JPB, such as punishment, reparation, reconstruction, and regime change. It is thus sensible, for example, that any punishment meted out for the war should be applied sparingly and primarily to leadership figures rather than ordinary soldiers, unless we are concerned with war crimes, other than the waging of the war itself. I should stress that I believe we can approach questions of blame in a cautious spirit without appealing to such doctrines as Michael Walzer's "moral equality of soldiers." (I have argued the case against Walzer's doctrine elsewhere and will not repeat it here (Coady, 2008a, 124–8).) Let me turn now to the areas of concern that I sketched earlier.

How to End a War Justly

Though much damaged by the insistence on "unconditional surrender" in the final phases of World War II, the path of negotiation is surely the primary ethical route to the ending of wars. The problems with unconditional surrender are well known, and I argued them in my book (Coady, 2008a, 272–5). Walzer also presents a strong case against unconditional surrender, but makes an exception for Nazi Germany because he thinks that the conspicuous evil of the Nazi regime "rightly places Nazism outside the (moral) world of bargaining and accommodation" (Walzer, 2006, 113). I agree that Nazism was a grossly immoral ideology and that leading Nazis and many followers deserved the epithet of "evil" but the real question concerned the prospects of negotiating with elements in Germany that were

---

4   One point of contestation is that President Obama says that the remaining troops will be withdrawn in 2011, but even here there is flexibility since a request by the Afghan Government for troops to stay will be seriously considered. Given the corruption in the government and its feeble authority in the country at the time of writing, such a request is highly likely to be made and probably accepted.

prepared to ditch Hitler and his cohort. The concerted refusal even to consider such prospects was one of the defects of the unconditional surrender policy, and the attitudes behind it, as applied to Germany. There were, of course, serious, possibly insurmountable, problems of detail in any such negotiations, but a certain undifferentiated demonization of the whole of "Nazi Germany" was probably a factor in this. Demonization of specific enemies and adversaries is a worrying enough element in international relations, as the effects of Anthony Eden's demonization of Egypt's President Gamel Abdel Nasser in the Suez affair of 1956 demonstrated, but there are even more damaging consequences of demonizing a whole nation or substantial portions of it. It is partly group demonization that supports such harmful political slogans as "no negotiation with terrorists," a slogan that is, in any case, eventually more honored in the secret breach than the public observance.

A central aim in the ending of a war for those who are fighting a just war should be that of leaving the peoples of the surrendering enemy nation in a position to contribute to an independent political life for themselves after the war. Such a political life may have to be very different from that which they had before and during the war, especially where its forms contributed to the waging of the unjust war. There will be many ways of achieving such a fresh political existence depending upon such matters as how much the current leaders of the enemy nation or group have contributed to war crimes or been driven by ideologies the persistence of which are likely to seriously inhibit prospects for enduring peace. But legitimate war aims should not include an occupation or dominance of the defeated enemy that would amount to ongoing colonisation or the imposition of an extremely harsh reordering of life, as was envisaged in the Morgenthau plan for Germany after World War II. This in turn means that what Kekskemeti calls a "vacuum" policy must be highly dubious (Kecskemeti, 1958). This is the policy that, to some degree, accompanied the unconditional surrender doctrine in World War II and involved the refusal, at least in theory, of any recognition of even temporary political authority in the losing side, even if it were allowed to reside in those who were opposed to the existing war leadership or had not taken a major role in that leadership.

The examples of West Germany and Japan are endlessly cited favorably in contemporary debates about the virtues of imposed "reconstruction" following unconditional surrender, Brian Orend, for instance, calling them "quite stellar and instructive examples" (Orend, 2002, 50). Nonetheless, there are reasons to be sceptical about the idea that the vacuum policy was a major contributor to the successful emergence of those two societies from the ashes of defeat. In fact a recent review of the economics and politics of post-war reconstruction, Christopher J. Coyne's *After War: The Political Economy of Exporting Democracy*, argues that in the case of Germany, "ultimate success was achieved not because of the occupation but despite it" (Coyne, 2007, 130). In the case of Japan, Coyne argues that MacArthur's genuine success was due, in large part, to the fact that he did not institute a military government, as in Germany, but utilized existing governmental institutions (such as the Diet, the Japanese parliament) and relied to a considerable extent upon indigenous actors (Coyne, 2007, 120). To this extent, the idea that unconditional surrender and total externally-driven reconstruction are vindicated by the examples of Germany and Japan is at least a considerable over-simplification.

In practice, of course, the lengthy occupation-cum-colonial rule in both West Germany and Japan of necessity required reliance upon local forms of political and economic leadership, even when it had ambiguous relationships to the occupying power. From the beginning in the case of Japan, although the Emperor's powers were reduced to the symbolic, General MacArthur's dictatorial power relied heavily upon the emperor's supporting role, since the emperor retained a huge amount of respect and personal authority with most Japanese. In West Germany, prominent political supporters of the defeated Nazi regime were brought into power as the occupation proceeded, and in Japan huge sums of CIA money were poured into the hands of politicians who had been prominent militarists but were now conveniently pro-American. This enabled them (as was intended) to set up a powerful political party that would be acceptable to the United States. The economic reforms in Germany that revived its economy and turned it eventually into an economic powerhouse were actually initiated by Ludwig Erhard, without the knowledge and against the authority of the occupying power, an authority that he fundamentally ignored (Coyne, 2007, 132).

Nothing can guarantee that leaving indigenous leaders, institutions and personnel in place and striving to promote genuine local self-government will not be risky, but even the total occupation of the enemy and political colonization for a time must face similar risks when the "reconstruction" of the losing nation is under way (witness, as already mentioned, the large number of Nazi officials who re-emerged in positions of power after processes of democratisation and de-Nazification) and it faces the additional burden of handling the chaos, suffering and devastation that a vacuum policy and a military occupation commonly create. In addition to the merit in preserving some indigenous political authority and structures, it is also important to try to preserve elementary civilian infrastructures in the defeated enemy territory during the conduct of the war. Destruction of communications facilities, electricity grids and other sources of power during the war are likely to make the transition to a peaceful world more difficult. (It may well in addition involve violation of certain principles of the JIB relating to the vexed question of collateral damage, but, even where such destruction does not violate the JIB, it may be prudent to restrict it in the interests of JPB).

There are of course several different scenarios for negotiated settlements that complicate how we should think of them in moral terms. I have considered the scenario of a side that (mostly) is engaged justly in the fighting of a war and has achieved sufficient dominance to make negotiations feasible on terms that respect its legitimate cause and the associated war aims. Here, the negotiations of the superior power should be informed by both principled and prudential considerations of the sort mentioned above. If we shift focus, however, and consider the dominance of an unjust side then other considerations come into play some of which I will consider later. (One that I have discussed elsewhere, but do not have space to revisit here concerns the dilemmas faced by the just side when it is certain to lose and seeks to negotiate an end to the war on disadvantageous terms (Coady, 2008a, 277)).

Punishments and Reparations

There is a tendency in traditional just war writing to view the waging of a just war as a form of quasi-judicial punishment for wrongs committed. For many reasons, it is better to avoid this model of war-fighting, as for instance the predominantly self-defence model does. Here there tends to be a conflict between two versions of a "domestic model" for providing moral justification for war fighting. One corresponds to that principally used by Walzer which sees the legitimacy of war as a form of law-enforcement akin to the law enforcement internal to states; the other sees the legitimacy residing more in the morality of individual self-defence (or defence of others) as would be licit in a state of nature, or more specifically in the absence of state protection for the endangered individual. Both of course have their problems, and there are relations between the two, but the more judicial flavour of the first can encourage the idea that a war is a punishment of the unjust enemy in some stronger sense than is required for effectively repelling his attack and ensuring a reasonable degree of safety from further harm. On either view, the practical problems of punishing the unjust enemy are such as to make any idea of punishment that might sensibly and morally apply to the defeated nation a thin one, mostly governed by an idea of security and some degree of reformation rather than retribution. There are several importantly different aspects of any such punishment. One concerns punishment of enemy individuals for the trio of Nuremberg offences—the crime of unjust war itself, specific war crime violations of the JIB, and crimes against humanity; another concerns punishment of the nation itself by deprivation of its territory, confiscation of property, limitations to its sovereignty (such as vast reductions in the military capacity allowed it), and drastic reconstitution of its economy. The question of reparations also may be considered here because it fits into a judicial picture and can be brought under the rubric of punishment.

Reparations have a long history in the aftermath of war, but this does not mean that they should form part of the JPB. There are several reasons why punishment of the nation, primarily in the form of reparations, itself should be treated with caution. The first is that it tends to punish those who bear little or no serious moral responsibility for the unjust war their leaders waged. This is particularly true of such measures as annexation of territory, removal of population, partition, long-term subjugation of sovereignty or promoting the dominance of favored ethnic or religious groups. It is also true, less dramatically, of reparations that require a tax burden falling on guilty and innocent alike. In addition, measures of this sort are likely to have bad consequences. Notoriously, the Versailles Peace arrangement at the end of World War I involved not only crippling reparations, but annexations of German territory both European and Imperial, all of which have been argued to have created the conditions of resentment and humiliation exploited by Hitler. (The central significance of this factor has been disputed by some scholars, but the plausibility of the claim is enough to illustrate my point.) In the wake of World War II, the Western Allies sought reparation by the seizure or destruction of German intellectual and real property, including industrial equipment and business assets, though they cancelled any further reparation attempts in 1952 (*Columbia Encyclopedia*, 6th edition (2001–7), entry on Reparations). Reparations also pose a major problem in that the victors determine the scale and type of reparation, and even just victors are unlikely to be paragons of impartiality when it comes to determining a share of spoils.

Walzer has tried to justify at least reparation by arguing that it is a legitimate case of "collective punishment" since "citizenship is a common destiny, and no one, not even its opponents (unless they become political refugees, which has its own costs too) can escape the effects of a bad regime" (Walzer, 2006, 297). I am not sure whether this is intended as a moral justification built upon an idea of common responsibility, or merely a factual claim about inevitability. Since Walzer rejects the idea that citizenship entails individual responsibility or guilt for the unjust war, then it seems more likely that his argument is of the form that since reparations are "surely due" to the victims of aggressive war and they can only be collected by taxing the defeated citizens of the enemy state, then those citizens must pay up. But there seem to me to be two flaws in this argument. One is that the existence of a legitimate compensation claim does not entail that it must be satisfied by whomever can be found to pay it. What is needed is some connection of moral significance between the wrong to be compensated and those who are to pay it, and talk of common destiny hardly establishes that. It is not clear to me what could establish it in the case of all enemy citizens. Perhaps, if the distributed costs of reparation were low enough, and a level of weak responsibility (or as I have elsewhere called it, accountability) were established for those citizens who did not oppose the war or were indifferent to it, then that might suffice, but it would still leave an illegitimate burden upon those citizens who had no such responsibility for the war, and there would be difficulty establishing just who bears what liability anyway. Second, I agree with Orend's argument (echoed by Bass) that there are ways that responsibility for reparations could be sheeted home to groups and individuals who were fully responsible, such as those who had profited handsomely and knowingly from the waging of the war or matters directly associated with it (Orend, 2002, 48; Bass, 2004, 408–9). Commonly enough, there are plenty of these even in a defeated country. Of course, there are complex and puzzling issues about the sort of communal responsibility that may be involved in governmental decisions more generally; there are many governmental uses of tax money that a citizen of a democracy or reasonably decent state might rightly object to without being thereby entitled to refuse payment of taxes, or part thereof. This may be explained by the idea that in a half-way decent state the taxation system usually serves good purposes, and is so complex a mechanism that a right of individual refusal would be very difficult to accommodate and would probably do more harm than good. This presumption may be defeasible in certain conditions, as the Berrigan brothers tax refusal over the Vietnam War indicates. The above considerations certainly suggest that reparations imposed on the whole community should be approached in a cautious if not sceptical spirit.

As for punishment of individuals for the crime of war or for war crimes, I agree that ideally perpetrators of war crimes should be punished for their violations of the jus in bello. But there are issues of possibility and feasibility in doing so when such violations have been authorized and carried out on a massive scale. Some of the most egregious perpetrators of the Nanjing massacres by Japanese troops were indeed executed at war trials after World War II, but the prosecution of the thousands of Japanese troops guilty of such crimes was probably unrealistic and would, in any case, probably have been counter-productive for the prospects of post-war peace. The lessons of truth and reconciliation commissions for situations of reconstruction after civil wars, revolutions etc are clearly relevant here. Many find the decision not to prosecute, or conventionally punish, war criminals, torturers or murderers under certain conditions a denial of justice, and indeed, in some situations, it is certainly that. In those situations, it would appear that the requirements of justice have been overridden by the requirements of peace, and the reconciliation of former enemies in the interests of order and future national development. It is, however, arguable that in some cases justice has not been ignored but merely (though drastically) modified since the miscreants are required to confront their victims, or relatives, confess their crimes, apologize and seek forgiveness.

This process itself could be viewed under the recently coined heading of "transitional justice" used to discuss transitions (whether peaceful or violent) from a form of oppressive government to a legitimate, usually democratic, one. It applies, for example, to the transitions in Eastern Europe from communism to democracy and in South America from dictatorships to democracies. At least one eminent theorist, Jon Elster, has argued that such transitions should avoid any recourse to punishments, restitutions or reparations because of the difficulties in establishing responsibility, rather than because of the need for peace (Elster, 1995, vol. 1, 566–8). I have no space to discuss Elster's argument in detail, but its scope seems to me unconvincing, partly because it both strengthens and weakens the ideas of harm and responsibility at different points to suit his purposes. Nonetheless, his argument does indicate some of the main difficulties in determining responsibility in post conflict or regime change situations, and provides cautions against zealous extension of a responsibility net. Apart from complex issues of responsibility that encompass matters to do with coercion, ignorance and so on, Elster is also concerned with the unfairness that resides in punishing conspicuous, if often minor, offenders, such as registered informers in former communist regimes, while not being able to punish ordinary members of the communist party who were expected to inform, and probably did so regularly, but were not registered (Elster, 1995, vol. 1, 566). This argument invokes both pragmatic and justice considerations, and it is a moot point whether ability to punish some offenders while being unable to punish others equally guilty always counts decisively against punishing any. Whatever about this, the pragmatic adjustments to justice in the interests of peace, truth and reconciliation are related to the various ways that normal judicial proceedings modify justice to take account of pragmatic circumstances, such as guilty pleas, willingness to inform on others, the overcrowding of prisons, and so on. In the case of peace and reconciliation, a good deal turns on just how successful such accommodations have been, and will be, but accommodations of this sort are clearly relevant to the post bellum conditions of international wars.

Beyond these considerations, an important argument, where prosecution is considered appropriate, concerns the need to prosecute impartially the major violators of the JIB on all or both sides of the conflict. This was a conspicuous failure of Nuremberg which made it appear "victors' justice" that was blind to palpable war crimes on the Allied side. The development of the International Criminal Court offers hope that such even-handed justice might become available in the future, although the failure of major powers, such as the United States to sign up to the Court is a disappointing impediment, especially as a primary US reason for rejection is that its military personnel might be prosecuted in the event of war. As the Crimes of War project puts it:

> For the United States, withdrawal from the treaty gives it a freer hand to launch a diplomatic offensive to minimize the chances that US military, governmental or other official personnel might ever appear before the court. On the same day that the withdrawal was announced, the government said that it was designating an official named Marisa Lino to negotiate bilateral agreements with countries to prevent them surrendering US agents to the court's jurisdiction.[5]

In the matter of "victors' reparations," some progress may be made by putting a United Nations agency in charge of considering claims for reparation both from individuals, companies and governments, as was done after the first Gulf War by a United Nations Commission on Reparation.

The problems with post war punishment are compounded when we shift focus and consider not what ought to be allowed to the just victors, but what are the best rules to have for victors generally, whether just or unjust. I shall revisit this issue later.

---

5   The Crimes of War Project http://www.crimesofwar.org/onnews/news-us-icc.html, paragraph 6.

Post-war Repair, Reconstruction and Governance

It may seem paradoxical that just victors should acquire obligations to restore the circumstances of the unjust, defeated enemy, but several considerations support this. One is that, as argued earlier, very many enemy citizens will not be seriously responsible for the war and deserve help in recovering from the distress of it; another is that where the enemy country has been attacked, much collateral damage will have been created to civilian property, health and livelihood, and even where this was licit (sanctioned by double effect or whatever) it is damage that should be repaired, preferably by those who inflicted it. Of course, just victors also have obligations to their own citizens who have been damaged by the war and any help they give to the defeated enemy must be balanced against that. Moreover, it should be expected that other nations, not involved in the war, have some moral obligations to help, just as they do with natural disasters. In addition to these considerations, the requirement to build what Basil Liddell Hart once called "a better state of peace" is recommended both by considerations of prudence and by the ideal of peace. Such a better state is likely to reduce the prospects of renewed war-making by the states so repaired.

Political reconstruction is partly covered by these considerations, but raises other questions. As mentioned earlier, the just conquerors should aim to put the people of a defeated nation in a position to contribute to an independent political life for themselves after the war and to play a reasonable part in international relations. Such a political life may have to be very different from that which they had before and during the war, especially where its forms contributed to the waging of the unjust war, but the difficulties of outsiders imposing such forms are considerable. The case for this is principally to do with the need for a stable, peaceful society that will be unlikely to undertake war again, at least in the medium term. The case against is that considerations of this sort can lead to the imposition of governmental arrangements that constitute the defeated state a mere puppet of the conquerors and subjugate their people to an oppressive regime. An extreme case of this is the reconstitution of the conquered East European states after World War II by the Soviet Union which was, broadly speaking, waging a just war of self-defence against Nazi Germany and its allies in the East. (It is plausible that the final stages of the Soviet push to Berlin was an unjust phase of that war, but this is only marginally relevant to the point I am making here.) The Soviet restructuring guaranteed that the Soviet Union would not be threatened by another war from the East, but in a way that imposed an unwelcome tyranny upon the peoples of East Germany, Poland etc. Even democratic victors can impose democratic forms, but keep such a firm hand on economic, military, and political factors as to make the result unsatisfactory from the point of view of reasonable self-governance. Israel's control over Palestine can be criticized from this point of view, since even when democratic elections are allowed, it seems that the Israelis and their closest allies will not allow them to work when the results are unpalatable to Israeli interests, as shown by the response to the electoral victory of Hamas.

Shifting Perspective

The case of Israel and the Palestinians is an interesting one because of divided opinion about the legitimacy of Israel's war that acquired the occupied territories in the first place. Whether justly or unjustly acquired, however, it seems that the lengthy aftermath has failed to meet some of the basic requirements of the JPB and this indicates the need to consider the JPB from both the perspective of just and unjust victors. Here, we should reflect upon the most appropriate rules and considerations for the end phases of what Vattel calls "regular war" (Vattel, 1867, Book 3, Ch. 4, section 66). His phrase usefully points in two directions: first, to the fact that wars, just, unjust, or containing confusing elements of both, are a regular feature of the history of humankind, and second, that there is an important task to find acceptable rules to regulate and restrain the conduct of such frequently occurring and often disastrous episodes with less regard for whether their origins are just or unjust. In Vattel's terms, we should turn our gaze when considering "regular war" from the dictates of the "necessary law of nature" to the "voluntary law of nations" which must prescind from the justice and injustice of the conflict's beginnings and attend only to rules, equally applicable to all sides, for regulating the conduct of the war (Vattel, 1867, Book 3, Ch. 12, Sections 188–91).

Looking at the JPB from the viewpoint of just victors will indeed take us a good part of the way on such a journey, since victors will usually see themselves as justified warriors, whatever the true facts, so should be conscious of the need for a number of the restraints we have considered. This is all the more likely where they do have a just cause, and the injustice of their war consists in their violation of other conditions of the JAB. Nonetheless, differences will arise for those cases where it matters that the defeated side had just cause and fought justly. In particular, permissions to exact punishments and to conduct political reconstruction may need to be treated with very great caution. Operating with a sort of "veil of ignorance" we might suppose that it would be best if victors went light on punishments, both individual and national, and also restricted the scope of political reconstruction. Nothing is easy from this perspective, since if your war was really just, you may in the abstract be entitled to some latitude in imposing conditions on the defeated side that you would not want them to have if they had won the war. The principal moral to draw from this is, I think, that protocols for punishment, reparations, reconstruction, and military occupation should be geared towards less harshness and less confidence in outcomes than might be justified by righteous victory. This is supported by the fact that regime change and other forms of political reconstruction are much more difficult than occupiers suspect, and imposition has very little prospect of success as well as carrying the presumption of moral taint.

There is a simple and powerful thought that needs to be confronted at this point in connection with unjust victors. This is the thought that unjust victors have a straightforward moral responsibility, namely, to abandon their immoral enterprise and quit the scene. But it is morally important that unjust victors, even when they realise that they are in the wrong, should not be able simply to disown any further responsibility and walk away. The slogan, "You broke it, you own it," sometimes called the "Chinese shopkeeper's slogan" is often invoked in connection with US post-war efforts in Iraq, but it is seriously misleading. The United States government, its armed forces and its allies are largely responsible for breaking Iraq, but that doesn't mean that they "own" Iraq. Iraq still belongs to the Iraqi people. The most that the slogan can instruct us in is that the Iraqi problem creates responsibility for those who "broke" Iraq.

The issues here are related to, in part identical with, those I have elsewhere discussed under the heading of "extrication morality" (Coady 1989, 163–225). We may well face situations where recognized immoralities of our own, whether individual or group in nature, face us with moral choices that require persistence for a time in the activities that fall under the blanket condemnation of injustice in order to most effectively and justly extricate from the immoral mess we have created. Unjust conquerors can gain no moral purchase on the reconstruction, political reorganisation of the defeated side from the justice of their cause, since they have no just cause, or have ignored some other condition of the JAB that disqualifies them from being legitimate conquerors. The most basic requirement might seem to be that they should just go away. But morality itself will often require that they make efforts to repair the damage they have done. Unjust victors may be compounding their injustices if they devastate a territory and then simply go home, so from the point of view of restorative justice it would be good that some considerations of JPB address their situation. In the aftermath of an unjust war, the unjust victors may thus be under a moral obligation to engage in armed peace-keeping and further fighting in order to restore an order that they have principally caused to collapse. Whatever they do in this respect should, ideally at any rate, be informed by the fact that their war-making is indeed unjust in its beginnings and overall orientation, so that the prevention of attacks upon innocent civilians by warring factions or helping in the restoration of damaged infrastructure should not contribute to a covert promotion of illegitimate war aims. They may not indeed acknowledge that their war is unjust, but others who do should urge such a restrictive policy upon them.

The argument that unjust warriors must take responsibility for what happens after their victory is an argument that some who condemned the Iraq war as immoral will accept for maintaining Coalition forces fighting in Iraq. It is a recognition of something like this that helps to support Moellendorf's claim that the JEB is independent of the JAB for he argues that an unjust war (which lacks a just cause) may nonetheless be right to continue when stopping it would lead to collapse and chaos in the state being unjustly attacked (Moellendorf, 2008). Although I believe there is a place for extrication morality, indeed, that it is one of the more plausible explications of what are called "dirty hands" problems, such arguments nonetheless need to be treated with care since they are easily adapted to self-serving ends, and the empirical judgements about likely chaos and bloodbaths tend to be elusive and fragile. We all remember the argument that the United

States could not leave Vietnam because victory for the Viet Cong (and the North) would mean wholesale massacre for the people of South Vietnam. When America departed, massacres did not occur, though oppressive re-education procedures were imposed on those who had been allied with the foreigners, and hundreds of thousands of South Vietnamese fled the country. The subsequent peace in Vietnam has had many negatives to it, but there can be no doubt it has been better than continued war. Frustrated invaders have a strong tendency to sustain the validity of their persistence by illusions that their presence is the only thing preventing disaster or promoting certain important goods, when often the reality is that their continuing occupation after an unjust war is a primary factor in an ongoing, deteriorating mess. The best solution should involve a definite short-term timetable for complete military withdrawal coupled with steady interim troop withdrawals and feasible plans for co-operative measures aimed at stabilisation, reconstruction and the hope of peaceful outcomes. Of course there are uncertainties attached to this strategy, and the political and military leaders of victor nations are reluctant to adopt withdrawal plans that might jeopardize the status of their victory. Hence they tend to stress the need for the long-term prosecution of pacification, restructuring, empowerment of local forces and so on; we are familiar with this language in Afghanistan at the time of writing. Yet it needs stressing that there are risks in every strategy, and involvement in indefinite or long-term occupation, war against entrenched local resistance, and predictable and extensive loss of life will usually have much higher costs than the alternative strategy recommended above.

Bibliography

Bass, G. (2004), "Jus Post Bellum: Postwar Justice and Reconstruction," *Philosophy and Public Affairs,* 32:4, 384–412.

Coady, C.A.J. (1989), "Escaping from the Bomb: Immoral Deterrence and the Problem of Extrication," in Shue, H. (ed.), *Nuclear Deterrence and Moral Restraint* (New York: Cambridge University Press).

Coady, C.A.J. (2008a), *Morality and Political Violence* (Cambridge and New York: Cambridge University Press).

Coady, C.A.J. (2008b), *Messy Morality: The Challenge of Politics* (Oxford: Oxford University Press).

Coyne, C.A.J. (2007), *After War: The Political Economy of Exporting Democracy* (Stanford: Stanford University Press).

Elster, J. (1995), "On Doing What One Can: An Argument Against Post-Communist Restitution and Retribution," in Kritz, N.J. (ed.), *Transitional Justice: How Emerging Democracies Reckon with Former Regimes* (Washington: United States Institute of Peace Press).

Kecskemeti, P. (1958) *Strategic Surrender: The Politics of Victory and Defeat* (Stanford: Stanford University Press).

Milne, S. (2010), "The US isn't Leaving Iraq, it's Rebranding the Occupation," *Guardian* (Comment and Debate Pages), August 5.

Moellendorf, D. (2008), "Jus ex Bello," *Journal of Political Philosophy*, 16:2, 126–30.

Orend, B. (2002), "Justice After War," *Ethics and International Affairs*, 16:1, 43–56.

Orend, B. (2007), "Jus Post Bellum: The Perspective of a Just War Theorist," *Leiden Journal of International Law*, 20:3, 571–91.

Rodin, D. and Shue, H. (eds) (2008), *Just and Unjust Warriors: The Moral and Legal Status of Soldiers* (Oxford: Oxford University Press).

Vattel, E. de (1867), *The Law of Nations; or, Principles of the Law of Nature, Applied to the Conduct and Affairs of Nations and Sovereigns. From the French of Monsieur de Vattel* (Philadelphia: T. and J.W. Johnson & Co.).

Walzer, M. (2006), *Just and Unjust Wars: A Moral Argument with Historical Illustrations* (New York: Basic Books).

PART II
Humanitarian Intervention

Chapter 4
High-Fliers: Who Should Bear the Risk of Humanitarian Intervention?

Gerhard Øverland

Humanitarian interventions will often result in loss of military and civilian life. Although soldiers' deaths may thwart further humanitarian operations, there is a tendency to argue that soldiers ought to accept a certain level of risk in order to avoid harming and killing civilians. They should not, for instance, fly at altitudes that eliminate all risk of being shot down by the enemy but increases the risk of civilian deaths from imprecise bombing. Such considerations may seem particularly relevant with regard to humanitarian intervention aimed at protecting already suffering civilians from a corrupt or evil regime.

This paper explores two reasons suggesting that it is nevertheless the prospective beneficiaries of the humanitarian intervention—the civilians of the invaded country – that should bear the brunt of the risk rather than the intervening soldiers. First, prospective beneficiaries of humanitarian intervention may have reason to accept considerable risk to life and limb in order to rid themselves of their country's regime. There would be no need for humanitarian intervention if the country's government dispatched its duties to its citizens. Second, the intervening forces are there for the sake of others, i.e. the civilians of the invaded country. Their duty to intervene—if there is such a duty—should therefore be based on a principle of assistance, requiring only moderate costs on part of the interveners.

Preliminaries

Given the likelihood of its many unwelcome consequences, I take no stand as to what could justify humanitarian intervention, but plausible candidates are genocide and severe human rights violations. Provided that the humanitarian intervention is appropriately conducted, I will assume it is justified and resistance to it is not. The defending party has no just cause and therefore no right to resort to defensive force. I accordingly reject the widespread view that soldiers in armed conflict have an equal right to kill, and hold that warring soldiers who fight on the unjust side have no right to do so and no right to kill enemy soldiers.[1] What the defending soldiers should do when faced with justified humanitarian intervention is to lay down their weapons and go home.

Intervening soldiers are nevertheless likely to meet with resistance from defending soldiers, some of whom may identify with the regime and fight for its survival out of loyalty or self-interest. Others may fight under duress. Both groups might include soldiers who are at least partly morally excused. And although the defending soldier has no just cause to take up arms, his innocence should impinge on how he is fought.[2] Concerning the last Iraq War, for instance, there was arguably a morally significant difference between members of the Republican Guard and regular conscripts. While the former benefited from their service, many of the latter would have been acting under compulsion. This difference gives the intervening soldiers reason to treat them differently, and to accept higher risk to themselves in confrontations with the latter. For simplicity's sake, I shall assume the defending soldiers are of the regime-supporting kind, and disregard any partial innocence on their part.

Civilians—or non-combatants—in the country targeted for humanitarian intervention will relate variously to the current regime. Basically, they can be divided into three categories. Some will be indifferent and neutral to a continuation of a regime that neither benefits nor harms them (and will lack concern for those harmed by it). Civilians may also be supportive of the current regime because of shared values or simply out of self-interest; indeed, some will enjoy privileges. I will call this type of civilian "regime-supporting civilians." The third group stands in opposition to the regime and in favor of a regime change. Call these the "to-be-liberated civilians." It is for this latter group humanitarian intervention is mainly launched.[3]

These simple distinctions may need to be broken down further. For instance, we could divide the to-be-liberated civilians into subcategories, those who suffer personally and those who do not but nevertheless would welcome intervention (perhaps because they have concern for those who suffer). But although both groups support intervention and perhaps regime change, the risk they would be willing to accept to that effect is likely to be very different. For simplicity I will overlook such variations.

In most cases where use of armed intervention would be a permissible option, one group of people is in power and controls the others through intimidation and use of force. Given the serious implications of armed conflict, the number of citizens in favor of regime change will normally need to be very high, either across

---

1   On the standard view see Walzer (2006). This view have been criticized by some philosophers (McMahan, 1994; Norman 1995; Mapel 2004). For an interesting discussion of McMahan's position see Steinhoff, 2008 and McMahan's response (McMahan, 2008).

2   I have dealt with the issue of confronting innocent soldiers who fight an unjust war in Øverland (2006).

3   There might be a fourth group encompassing all the non-related civilians who are part of the conflict at all, as they may live in a different country altogether.

the country or in certain regions, in order for intervention to be permissible.[4] The latter would justify intervention in support of secession.

Proportionality is an essential aspect of determining the ethics of war, and in this context it has to do with identifying who should bear the cost of the intervention. Leaving to one side the question of proportionality between the intervening and defensive forces, candidates for trade-offs are basically the intervening soldiers versus the different categories of civilians in the invaded country (assuming that civilians belonging to the country of the intervening soldiers are not under any threat). Two situations will help illustrate the kind of considerations we might need to take into account. One is the above-mentioned altitude issue, where the bombing of military installations from high altitudes increases the likelihood of civilian deaths, while bombing from lower altitudes increases the risk to the lives of the aircrew. The second situation involves attempts to take over cities and residential areas, where the risk of civilian harm is traded off against the risk to intervening soldiers.

Civilians

Granted that the intervening soldiers comply with standard rules of engagement, they take care to limit civilian casualties. On occasion, however, civilians will die as a consequence of the intervention, and the question thus arises as to the permissibility of acting in ways which knowingly puts the lives of non-combatants at risk. If required by military necessity it is generally accepted that it might be permissible to kill a limited number of civilians as a corollary of killing soldiers or destroying military installations. What number it is permissible to kill is hard to pin down. According to the proportionality condition the means used to prevent an attack can be permissible only if they are proportional to the interest at stake.[5] But there is substantial controversy as to what should be factored in to determine the number of civilians that may be killed to advance an important military goal, and in particular whether the civilians' (partial) responsibility for the war in the first place should impinge on their immunity. It seems anyway difficult to deny at least

---

4   It is generally agreed that an intervention is not morally permissible unless the intended beneficiaries consent to their rescue. Richard Miller writes, "Outsiders ought to have warranted confidence that the vast majority of the intended beneficiaries of the intervention consent to the risks on the basis of adequate information" (Miller, 2003, 224–5).

5   For a discussion of proportionality, see Hurka (2005). According to Hurka the "proportionality condition says the collateral killing of civilians is forbidden if the resulting civilian deaths are out of proportion to the relevant good one's act will do; excessive force is wrong" (p. 36), and he maintains that "the condition allows bombing a vital munitions factory if that will unavoidably kill a few civilians, but forbids killing thousands of civilians as a side effect of achieving some trivial military goal" (pp. 36–7).

a limited permissibility of killing civilians as a side effect unless one wants to deny the permissibility of most modern military operations.

One might want to take civilians' responsibility for an aggressive war into account.[6] Civilians may be culpable causes of the unjust threat their soldiers pose to another party. Their actions may have stirred their government to wage an unjust war. They may have written inflammatory articles, joined public demonstrations, and, most common in democracies, voted for the government that is now waging war on their behalf. These civilians' weight in the proportionality calculus should be discounted according to their moral responsibility for the war. To illustrate how this discounting is supposed to work, assume that we need to bomb a munitions factory. It might now be permissible to proceed even though doing so will kill ten civilians as a side effect when these civilians can plausibly be said to have some moral responsibility for the war, while not permissible if all ten were innocent. However, even when the civilians were innocent it could be permissible to proceed if only five of them were in the neighbourhood. The thought here is not that the additional five *deserve* to be killed or anything of that kind as a consequence of their partial responsibility. The idea is that if we think that killing five civilians as collateral damage is permissible in the first place, we must already have accepted that their immunity is not absolute, and then a plausible suggestion is that their claim on protection is further reduced depending on their level of moral responsibility for the ongoing war.

It is important not to think of immunity in black and white terms, but as something that comes in degrees. We might therefore prefer to talk about *civilians' claim on protection*, and not about their immunity. With regard to humanitarian intervention, regime-supporting civilians may be morally responsible for helping the repressive regime to stay in power, or at least for failing to withdraw their support. And just as civilians may have their claim on protection discounted by supporting unjustified aggressive warfare, regime-supporting civilians' claim on protection could be reduced both as a consequence of their support of the corrupt regime and as a consequence of their (informed) acceptance of benefits and privileges.[7]

Clearly, this could not be said about to-be-liberated civilians. They have not sent soldiers anywhere and do not receive benefits or privileges from a corrupt regime. They are therefore not culpable causes of the harm done by their government, and cannot lose their immunity on the basis of any responsibility-based account. Hence, whether or not we think civilians should have their claim on protection discounted depending on their responsibility for the danger posed by unjust soldiers, or for being supportive beneficiaries of a human-rights-violating regime, none of these applies to to-be-liberated civilians. One might therefore want to conclude that they

---

6   See, for instance Mahan (1994); Øverland (2005a).

7   Perhaps even the indifferent civilians may be said to share some moral responsibility, as they ought not to be indifferent with regard to severe suffering on the part of other people.

have the strongest possible claim on protection. But that would be to conclude too quickly.

Acceptance

There is a tendency to assume that a higher rate of collateral damage is justified in ordinary defensive wars than may be justified during a humanitarian intervention. According to George Lucas, for instance, "military personnel engaged in humanitarian actions may not be entitled simply to protect themselves at all costs from harm, or even to inflict unintentional collateral damage on non-military targets or personnel by the principle of double effect, even though both of these eventualities are commonly excusable under the "war convention" in traditional combatant roles" (Lucas 2003, 78). It is not obvious that we should agree with these claims.[8] In fact, a closer look seems to suggest that the contrary follows from a standard view of the permissibility of using defensive force.

In a defensive war the civilians of the opposing side are not initially under threat. They are first put in danger by the defensive side's choice of engaging in defensive action. The duty not to harm them is therefore based on the duty not to harm others when taking defensive measures. There are widely accepted limits to the permissibility of the use of force under such circumstances. Imagine that Tom attacks Mary, and that the only way she can save herself is by killing him with the flamethrower she happens to have brought with her. It would not be morally permissible for her to do so if proceeding involved a 50 percent risk of setting fire to the house behind Tom and killing the people inside it. Although Mary may have a just cause, namely defending herself against an unjust aggressor, her act is impermissible because it has the death of other people as a likely side effect. It is in general not permissible to use defensive force indiscriminately, i.e., when it affects third parties in addition to aggressors. This must be true with regard to war as well. Killing civilians as a side effect would require proportionality between the military goal and the number of civilians killed (which may or may not involve an evaluation of their responsibility for the war in the first place).

By contrast, in a humanitarian intervention the to-be-liberated civilians are already under threat and are therefore not really third parties. The intervention is justified exactly because they are under some sort of threat from their own government. The intervening soldiers' duty not to harm them should accordingly be informed by the to be-liberated civilians reasons to accept risk for the purpose of seeing a successful intervention.

Let me illustrate with the example of Tom and Mary. Suppose that Tom does not attack Mary this time, but that he holds four people hostage in a house behind him. Mary can intervene, but her only way to do so is by killing Tom with the flamethrower she happens to have brought with her. Unfortunately, it is not easy

---

8   I assume that Lucas here uses "excusable" to mean "justifiable."

for her to focus the flames directly towards Tom; there is therefore approximately fifty percent risk that the house will go up in smoke as well. The permissibility of Mary to proceed in this case would depend on whether it is plausible to assume that the people in the house would prefer her intervening by using the flamethrower or rather face the continued risk associated with being hostages. Since the risk associated with intervention is very high, the situation of the hostages would, of course, need to be pretty desperate, but the proposal is nevertheless that when evaluating the permissibility to proceed one would need to take such considerations into account.

Now, given that the hostages—and by analogy the to-be-liberated-civilians— are under severe duress, one might wonder whether we should really take their willingness to accept risk as a genuine form of consent. To evaluate that we need to consider their interests. The hostages' most important interests at this point seem to be that their risk of harm and death is reduced. This is why even a risky intervention might serve their interests when the alternative is an even more precarious status quo.[9] Moreover, being in an unfortunate situation does not normally count as disqualifying people from making informed consent, a patient choosing between a risky operation and severe illness being a typical example. This person may be best served by a hazardous surgical procedure. As for situations of duress, it seems that such situations render consent invalid only when those posing danger by extortion and those getting the benefit of the consent are one and the same. Hence, any consent made by the hostages to the demands of the hostage takers would count for nothing.

The to-be-liberated-civilians (or the hostages) may, of course, also have an interest in deciding whose hand exposes them to risk. They may perceive a difference in being harmed as a consequence of an armed intervention as compared to as a consequence of their own government's tyrannical rule. A slight reduction in risk may therefore not suffice. Yet, while some people might perceive it as more objectionable to be harmed during an intervention, others could prefer to be harmed in a failed rescue mission rather than at the hands of their government. It is not obvious that we should assume they have an interest either way. Having said that, we do need to take into account the possibility that as the consequences of the to-be-liberated civilians' willingness to bear risk start to materialize they may change their mind and even turn against the intervening forces. Consequentialist

---

9   An analogy is found in the infamous example introduced by Bernard Williams (Williams and Smart, 1973; Williams, 1998, 34). Being a visitor in a South American town, Jim is given a choice by Pedro to kill any one of 20 men who are lined up against a wall. All of the men will be killed unless Jim agrees to kill one of them. To identify a reason for Jim to choose to kill one we can note that the initial 20 probably would unanimously agree to the outcome in which one of them were randomly killed. At the point at which Jim is introduced into the situation all the twenty people are faced with the prospect of being shot, and we have every reason to believe that they would prefer to see one of their number randomly chosen and killed in order to rescue the rest (Øverland, 2005b).

considerations against killing to-be-liberated civilians must therefore be taken into consideration as prolonged exposure to risk may turn them against the intervention. And this may be the case even though they had reason to prefer intervention with that risk.[10]

None of these considerations make for easy judgements, and we do not really have the option to ask whether individuals who are supposed to benefit from the intervention are willing to accept the required risk. Nevertheless, considerations about the to-be-liberated civilians' willingness to accept risk ought to guide our evaluation of the permissibility of any humanitarian intervention, and since we cannot run a census, taking into account what we can plausibly assume as being in their interest seems to be our best option.

In defensive wars, the military goal of the defensive army is likely to be in direct opposition to the interests of the civilians on the other side. The permissibility of defensive military operations that risk civilians' life can therefore only appeal to these civilians' responsibility for the aggressive war—something which is both controversial and empirically difficult to verify—in addition to standard considerations like military necessity and so forth. With regard to a humanitarian intervention, the situation is different. The military goal of the intervening army is very much in line with the interest of the civilians who are in danger of being harmed. Not to take this fact into account misses an essential point.

Just as those in the house may welcome an end to Tom, people might want to see their country's government overturned. The civilian population of a nation may accordingly have good reason to accept a certain level of risk if it means getting rid of the government. These civilians may have a lot to win by a humanitarian intervention; it is after all essentially done in order to help protect them from a bad regime. In such cases, if it means putting to-be-liberated civilian lives at risk, it could be permissible because it is in their interest that we do so (given the alternative). The basic argument is simple enough. *If the to-be-liberated civilians have reason to accept a particular risk of being harmed and/or being killed in order to be liberated from their oppressive regime, intervention not exceeding that risk would not wrong them, provided that the interveners do not have a duty to do bear risk for them* (to be discussed in the next section).[11]

Note that the acceptance-based argument for reducing civilians' claim on protection seems to apply to children as well. This is in sharp contrast to the responsibility-based account in defensive wars. Children cannot seriously be held responsible for sending soldiers anywhere. Children may, however, have an interest

---

10   As mentioned above, there are likely to exist civilians who are supporters of the regime. Their claim on protection has been reduced by a responsibility-based account. Some of them, however, perhaps contrary to what they think themselves, are likely to benefit from a regime change. This is not a problem, however, as it will just add to the reason for reducing their immunity.

11   In this way one could believe that invading Iraq in 2003 was justified as an intervention aimed at regime change, though not on grounds of self-defence.

in seeing their current regime substituted with another. They, or their guardians on their behalf, could accordingly have reason to support an intervention even at a risk of being harmed or killed. Hence, in contrast to a responsibility-based argument for killing civilians, the acceptance-based argument would apply to children and adults alike. Moreover, the acceptance-based argument for a reduction in the claim on protection could even cover children of regime-supporting adults. Even they, or more plausibly guardians on their behalf, would have reason to support a regime change that allowed for future flourishing.

When considering consent the badness of the current regime will arguably impinge on the permissibility of intervention. If present conditions are very bad, the repressed people have reason to accept a higher risk in order to be rescued from the oppressors. The acceptance-based argument simply indicates the highest risk it would be permissible to impose on the to-be-liberated civilians. As a consequence, if no duty to intervene applied to the intervening forces, intervention at the highest risk level would not wrong the to-be-liberated civilians.

But it is not always permissible to impose on the to-be-liberated civilians the maximum risk they are willing to accept. This is because the level of danger facing the to-be-liberated civilians will not only influence the level of risk they are wiling to accept, the badness of their situation will increase the duty on part of the intervening forces to shoulder cost when coming to their rescue as well.[12] In the simple example, whether or not Mary is permitted to intervene in a particular way is not determined solely by the willingness of the hostages to accept risk. Ideally, if she could chose between different options, she should pick the one exposing those in the house to as little risk as possible. Yet, if the different options would require her to bear different levels of risk, it is not obvious that she ought to pick the one imposing as little risk as possible on the people in the house. There are limits to the risk that we can require Mary to take, even when faced with saving the life of many others. Suppose, for instance, that Mary's other option to intervene would involve sneaking up to the house under cover of darkness and releasing the hostages. It might be a honorable choice on her part to do so, but being a very risky operation, it could perhaps not be required from her. And if not, she would be permitted to use the flamethrower from afar as a second best.

Likewise, the duty on part of the intervening forces to shoulder risk in order to reduce risk to the to-be-liberated civilians could make it obligatory on their part to intervene in a way that exposed the civilians to less risk than the maximum level the to-be-liberated civilians would be willing to accept. This is uncontroversial; the intervening soldiers ought to comply with whatever duty that applies to them. The question is what duty does apply to them.

---

12   I am grateful to Paolo Tripodi for pressing me on this point.

Soldiers

Intervening soldiers' claim on protection may be influenced by factors pulling in different directions. Insofar as we are talking about humanitarian intervention for the benefit of people in a foreign country, we might think the responsibility for engaging in such activities should be dictated by a principle that only requires the intervening parties to bear very moderate risk. On the other hand, not unlike policing, invading soldiers may need to accept a higher risk in order to protect civilians and facilitate arrest rather than liquidation. While I think there is much to the first point, the second seems to rest on unsound assumptions.

Assistance

The intervening party could previously have supported the corrupt regime and for that reason had some contribution-based responsibility to intervene. That would give the intervening party an increased responsibility to shoulder risk. Just as we think ordinary contributors are required to bear higher cost to help protect their victims from harm, the intervening party could be required to do likewise. Leaving that option to one side, and assuming that the invading army has neither contributed to the creation of the corrupt regime nor sustained its ability to stay in power, its duty to intervene – if there is any such duty at all – must be based on an application of a principle of assistance.[13]

The essential element when justifying humanitarian intervention is that some people are victims of unnecessary suffering as a result of a policy adopted by a particular regime. The main goal of the intervention is therefore to bring assistance to these suffering people, and, if necessary, facilitate long-term improvements by inducing a regime change. We may accordingly not be entitled to require the intervening soldiers to bear considerable risk. After all, and pace Peter Singer, it is, for instance, generally not required that you save a drowning child if it puts you at considerable risk. What is generally accepted is that you save a child when the cost to you and others is moderate. In "Famine, Affluence, and Morality," Singer argued that we have responsibilities to assist the global poor by alluding to an analogy of a person passing a shallow pond where another individual is about to drown. Just as the former bears responsibility for saving the latter, we have a responsibility to assist the poor. According to Singer, a plausible principle that would explain our reaction to the pond case, and which would also lead us to recognize our responsibility in the global poverty case, states that if we can prevent something

---

13   One of the main points of the newly established Responsibility to Protect doctrine is that there is not only a right to intervene in order to protect vulnerable people from serious harm; there is also a responsibility to do so when necessary (ICISS, 2001). Well and good, but the question of what cost the intervening forces have a duty bear still needs to be addressed.

bad from happening without sacrificing anything of comparable significance, we ought to do it (Singer, 1972). A less demanding principle, which would support the same judgements about the two cases, states that if we can prevent something very bad from happening to other people at moderates cost to ourselves and others, then we ought to do it. Needless to say, it is the less demanding version—if any—that is widely accepted.

Hence, the fact that it is a question of a humanitarian intervention for the benefit of a people in a foreign country may suggest that most people would think that we ought not to order soldiers to bear additional risk, let alone sacrifice their life, simply to lessen the risk on the to-be-liberated civilians. People may, of course, decide to support Singer's demanding version of the principle, or any in between, and adjust what they think soldiers ought to risk accordingly. However, this option would require substantial revisions of what people normally think individuals ought to do for others, in particular with regard to global poverty, as well as imply drastic changes to the distinction between doing and allowing harm. I shall assume that people are not ready to support any demanding version of the principle of assistance, and that it is a moderate version we need to consider.[14]

Most interventions would seem to place a higher risk on the intervening soldiers than people in general are required to bear in order to save others from harm. The very idea that we could have a duty to intervene is therefore peculiar. After all, the risk would have to be imposed on some of us, namely our soldiers. If we do not think people have a duty to shoulder considerable risk to assist people with whom they have no special relationship, not even when the latter are in dire need, it is hard to think we can have a duty to intervene at all if such intervention will impose any substantial risk on the intervening soldiers.

Most humanitarian interventions therefore appear to be supererogatory actions on the part of the individuals involved. One implication of this could be that whenever it is possible to intervene with different risk-distribution below the limit set by the to-be-liberated civilians' willingness to accept risk, it would be permissible for the intervening forces to take the less risky option, and not for us to require the soldiers to let the security surplus pass to the to-be-liberated civilians. This comes about because there may be various tactics of intervention allocating the risk between the to-be-liberated civilians and the intervening soldiers differently.

For simplicity, assume that the minimum necessary risk to the invading soldiers is 0.2; no successful intervention could be launched below that risk. But given their skills and ability to intervene they are able to launch a coordinated attack at this level. It would involve aerial attacks and bombing from high altitudes. As individuals, and given the present need of the to-be-liberated civilians, the intervening soldiers would have no duty to intervene with a risk above 0.2. As for the to-be-liberated civilians, the highest risk they are willing to accept is 10, which

---

14   See also Singer's latest book *The Life You Can Save* (Singer, 2009) and the review (Barry and Øverland, 2009).

also happens to be the risk they need to shoulder if the intervening soldiers choose the strategy that exposes them to as little risk as possible (high-flying).

Suppose furthermore that it is possible to reduce risk to the civilians, but that doing so would increase risk to the soldiers. There are equal numbers of civilians and soldiers. Given the nature of military tactics, it is possible to reduce risk to the former without increasing the risk to the invading soldiers by the same degree, thereby reducing overall risk to people. There are three ways to intervene: the first is high-flying, mentioned above, with a risk to the to-be-liberated civilians and the intervening soldiers of 10 and 0.2. Then there is the option of reducing bombing altitude, call it "low-flying," giving a risk of 5 and 0.4 respectively. As a third option one may abandon bombing from the air altogether and move in with ground troops only. This would reduce the risk to the civilians to 2.5 while increasing that to the soldiers to 0.8. Call this "no-flying."

According to the proposal so far, intervention by high-flying would not wrong the to-be-liberated civilians. The to-be-liberated civilians were willing to accept the risk, or given their situation, they would have reason to accept that risk, and the intervening soldiers would, as individuals, have no duty to bear higher risk than the one they are actually taking, namely 0.2. Could the intervening soldiers nevertheless have a duty to bear higher risk simply because they are soldiers, requiring them to pursue a low-flying or no-flying intervention strategy?

Policing

According to Michael Walzer, "humanitarian intervention comes much closer than any other kind of intervention to what we commonly regard, in domestic society, as law enforcement and police work" (Walzer, 2006, 106). Other scholars share this view. According to George Lucas, "the very nature of intervention suggests that the international military 'police-like' forces (like actual police forces) must incur considerable additional risk, even from suspected guilty parties, in order to uphold and enforce the law without themselves engaging in violations of the law." (Lucas, 2003, 73).[15] If this were correct it could change the rules of engagement, placing, for instance, a much higher burden of risk on the invading soldiers in order to protect defending soldiers and civilians alike. Police officers are not supposed to use lethal force unless special circumstances make it necessary.[16] In particular, they are not supposed to use deadly force at all if bystanders could be harmed.

---

15   It is not clear which violations of the law by the police Lucas here has in mind. But the question is not really whether they ought to take risk to avoid violating the law. Of course they should not violate the law. But to assume that exposing civilians to risk is to violate the law is begging the question.

16   See Lackey (2006). Lackey also states that "Soldiers are supposed to use deadly force, unless special circumstances argue against it." But while that certainly may be true

Leaving to one side the question whether there is sufficient relevant similarity between humanitarian intervention and police operations to justify the claim that the activities of the one should be guided by the rules of the other, there are a few hurdles with the police-based argument. Although it may seem straightforward that both the police and the intervening forces ought not to impose risk on innocent bystanders, I have already indicated why imposing a risk on to-be-liberated civilians may be permissible, namely that it is in their interest that we do so (when the alternative is no intervention). Moreover, these civilians are not really bystanders; bystanders are not supposed already to be under threat. To-be-liberated civilians are therefore more akin to hostages than bystanders in ordinary police work.

In hostage situations, the goal of the police is similar to the goals of the hostages. They all seek the liberation of the latter. In such situations the police, or whatever official body is in charge, need to reduce the risk to the hostages. If that requires intervention, and it can be executed without risk to third parties (bystanders), intervention would be the preferred option even when it imposes a considerable risk of harm on the hostages. It is after all their best option. In this respect, the to-be-liberated civilians are like hostages; they may benefit from intervention if it reduces their exposure to risk, all things considered. In neither case must intervention impose risk on innocent non-threatening third parties (unless that risk is very low), nor should such action be taken when there are less risky options available that will do a better job at liberating the people in need like negotiations or diplomacy.

We now need to investigate whether it is the case that whatever risk the hostages would be willing to incur in order to be liberated, it would be the prerogative of the police to secure for themselves the less risky option as long as they intervene with risk to the hostages that is below that limit. Of course, they would be required to bear whatever costs are dictated by the principle of assistance, as it applies to all individuals. The question is whether they would have a duty to shoulder higher costs than dictated by this principle. After all, that they have such a duty seems to be an implicit requirement of the police-based argument for the duty to bear additional risk on part of the intervening soldiers.

In an attempt to illustrate the dilemma let me continue to assume that any individual has a duty to bear a risk of 0.2 to save another person's life. The question is whether a police officer when confronted with a particular situation would have a duty to bear additional risk, 0.4 say. The answer may seem obvious, they do have such a duty, and the reason is that they have a role-based duty to do so. Hence, even in a hostage situation, when liberation would impose considerable risk on the hostages, the police must try to reduce that risk by taking more on themselves. If that is so, intervention as a form of policing could imply heightened protection of the to-be-liberated civilians and greater risk to the liberating soldiers.

---

within the traditional ethics of war, I think this assumption is not very well founded and trades on the idea of an equal right to kill.

But is it obvious? Do we really expect the police to incur considerable additional risk as compared to ordinary citizens in emergency situations? Well, at least we shouldn't do so when the person in need is about to drown. A police officer is not required to jump into dangerous water and take additional risk in order to save other people from drowning. If he or she has a duty to bear additional risk, it must be in situations that are police related in the sense that their special skills give them a particular advantage to deal with the situation. Often this means that there is a crime, or a potential crime involved, which is not the case with the drowning person.[17]

In general the police may have to do things citizens are not obliged to do. They have to drive around in their patrol cars, direct the traffic, look for suspects, and so forth. In short they have to go to work, and their job does involve a background of risk. And this risk may very well be higher than for other professions, like being a dentist or an academic. By volunteering to take the job, they volunteer to do what it consists of, and to carry the daily risk associated with it. The question I ponder is whether a police officer when confronted with a particular situation has a duty to bear additional risk.

It may be difficult to judge, as ordinary citizens are not really under any duty to engage in police related activities. They are supposed to call the police.[18] Let us nevertheless investigate the hostage situation alluded to above. But now suppose that there are two different hostage situations, one where the police are present, another with only civilians at hand. In both cases intervention seems to be the only way to save the hostages. Luckily, the citizens, who are not as good at doing such things as the police, are faced with a fairly easy task. They will accordingly be able to intervene successfully by taking on no more than 0.2 risk. The task of the police is more difficult, but they can also intervene successfully with a risk at 0.2 due to their competence in such matters. In both cases the risk to the hostages will be 10. However, both the ordinary citizens and the police may choose a second option reducing the risk to the hostages (to 5) by increasing the risk to themselves (to 0.4).

Would the police officers have a duty to pursue the option that imposes less risk on the hostages by taking more on themselves? We have already stipulated that individuals' duty to bear risk in such a situation is 0.2, and we have therefore no reason to blame the citizens for not taking the more risky alternative. Of course, the police are able to do things that ordinary citizens are not. But that is best explained by training, the issuing of special equipment to contain risk (body armour and firearms, etc), and a coordinated ability to act so that they can do much more at the same risk level in police-related operations. That is accounted for in

---

17   It could be the case, of course, as another may have thrown out the person now struggling to stay afloat. I am not sure how this would impact on the situation.

18   Note that they are not supposed to call the police because it is the duty of the latter to bear the risk, but because the police are better able to deal with police related situations.

the above story. The operation launched by the police is more complicated and ordinary citizens would have been unable to perform it at an acceptable risk.

I am not sure, but it is not at all clear to me that police officers would have to accept a greater risk of death or injury than other citizens in order to save the hostages. Return to the drowning child. One could perhaps think that a lifeguard would have a duty to jump into the water at a higher risk than ordinary people. But isn't it rather the case that this person is an expert swimmer with the necessary equipment, and therefore able to rescue people who get into difficulty in a swimming pool or at beaches at an acceptable risk level, that is, 0.2?[19]

This way of looking at it would jettison the need to appeal to role-based duties. Police officers, lifeguards, and fire fighters would accordingly have no duty to shoulder higher cost to protect people than ordinary individuals do. Clearly, in the long run members of these professions might be exposed to an aggregated higher risk than other people, and that could warrant compensation in form of early retirement or a risk surplus in payment. Anyway, the mere fact that we seem unable to answer this easily and judge with confidence that there is a significant additional duty on the part of the police to run risk indicates that the difference cannot be very significant. Hence, alluding to the rules of engagement for the police cannot easily be used to demonstrate that intervening soldiers have a duty to incur *considerable* additional risk.

One question remains. We sometimes think soldiers are permitted to kill civilians as a side effect of pursuing a military objective; may the police likewise kill bystanders as a side effect when pursuing a police related objective? The answer is, of course, that they cannot. However, that the police ought not to expose bystanders to risk in their attempt to enforce the law should not be confused with the question at issue, namely the extent to which they have a duty to incur considerable additional risk in order to avoid doing so. The mere fact that police officers ought not impose risk on bystanders when pursuing criminals doesn't reveal any interesting observation. This is simply a consequence of the fact that in most cases—pretty much all—the apprehension is just not important enough to warrant exposing bystanders to substantial risk. That doesn't imply that the police have to bear additional risk; they should simply retreat and allow the suspect to escape. By contrast, in a just war the overall goal of winning lies in the background as a justifying cause for accepting collateral damage.

Let me end this section by making a simple observation. Any police-like intervention would require an asymmetrical power-balance between the intervening and the defending army. We cannot even begin to contemplate the possibility of

---

19   An interesting question, raised by Sam Butchart, is what about the very competent person who is not a lifeguard; would she have a duty to do what a lifeguard has a duty to do? Well, not if a lifeguard is on the job. But if not, and she has the necessary equipment, which she normally would not, the ideas is that she has. The role, would not add to the duty to take risk in a particular situation. It would only make it the lifeguard's duty to intervene rather than another person's duty with the same skills.

intervening in accordance with rules of policing unless there is an overwhelming strength of force on the intervening side. Unless there is such an asymmetry in strength there will simply be no reasonable hope of success. Moreover, a look at death tolls suggests that soldiers engaged in intervention are exposed to much higher risk than the police. For instance, Norway has lost five soldiers in Afghanistan since 2003, despite the fact that fewer than 500 men are deployed there at any one time (most of the time the number is closer to 200). In the same period, one member of the approximately 12,000-strong Norwegian police force was killed on duty. It is therefore not clear what intervening soldiers would be able to achieve by applying rules of engagement used by the police.

The Goal of Intervention

There may be other, more direct, ways to defend the view that intervening soldiers ought to bear additional risk. The very idea of intervention as a life-protecting enterprise may seem to require that the intervening soldiers incur additional risk to limit casualties among the civilian population they aim to assist. Humanitarian intervention aims at protecting already suffering civilians. It has accordingly been argued that the intervening forces must be willing to incur risk and put themselves in harm's way for the sake of the moral ideals these forces aim to protect (Lucas, 2003, 93).[20] This argument appeals to the very mission of the intervening soldiers and attempts to ground their duty to accept risk in the humanitarian goal for which they fight.

The problem with this approach is that it is far from clear that this kind of duty applies in other instances of trying to safeguard people from danger. I have presented the liberation of hostages as a case to the contrary. In the more mundane rescue case where you jump into the water to save a drowning person, you might, when faced with panicked behaviour from the person you try to help, be permitted to cancel the rescue-mission and push the person under the water in order to extricate yourself if that is the only way you can get up on dry land again. And that is true even though you could have allowed the panicked person to save herself by pulling you under instead and to use you as a stepping-stone. The mere fact that you have ideals when entering into a rescue situation does not imply that you have to bear all the risk.

Of course, there are limits to the risk it would be permissible to impose on the to-be-liberated civilians, but we cannot simply assume that because some people

---

20   Paolo Tripodi suggests that "The key to begin to establish a proper peacekeeper's mind-set is to change and broaden the focus of his human rights training. What is important is to bear in mind that the peacekeeper ethos is significantly different from the warrior ethos. While the latter is ready to accept a high level of risk to protect his fellow citizens and the nation's interest, the peacekeeper subscribes to humanitarian values and he/she is ready to risk his/her life to protect any fellow human being" (Tripodi, 2006, 229).

volunteer to help assist others they have a duty to put themselves in harm's way come what may. Why should we assume that a person incur such a duty?

Lucas talks about "consistency" between the ends and the means (Lucas 2003, 78). But that just doesn't seem to do the work. We would then first need an explanation of why we do not require the same consistency with regard to ordinary rescue cases, and an explanation of why it is plausible to require it from intervening soldiers. Lucas does neither. As for the first, we clearly think it is permissible that we leave some risk to be shouldered by those we aim to rescue; the drowning person being just one out of many examples. It is hard to see that humanitarian intervention is altogether different, and it would at least require a substantial separate argument. To say that the intervening soldiers' business is to protect people from harm is not enough, as doing so may be done with various distributions of risk.[21]

Of course, the intervening forces are not supposed to engage in human rights violations, or any type of atrocities that the current regime commits.[22] But why should they? There is no reason to commit human right violations when intervening. To assume that exposing the to-be-liberated civilians to risk as being an instance committing the kinds of act they are intervening to prevent is begging the question. I have tried to demonstrate that exposing the to-be-liberated civilians to considerable risk may in fact be permissible. Moreover, and needless to say, the current corrupt regime is not exposing their population to risk in order to liberate them.

Concluding Remarks

In this paper I have indicated reasons for being sceptical about the assumption that intervening soldiers have a duty to bear considerable additional risk when engaged in humanitarian intervention. By contrast, civilians that are to be assisted or liberated in the intervention, have reason to accept additional risk of being killed as a side effect. At one point, however, I indicated that when the consequences of their willingness to bear risk start to materialize, the to-be-liberated civilians might change their mind and even turn against the intervening forces. Consequentialist considerations against killing to-be-liberated civilians must therefore be taken into consideration as prolonged exposure to risk may turn them against the intervention. One likely upshot of my argument could therefore be that we—as citizens of

---

21 Interestingly, Lucas' consistency claim seems to imply duties toward the global poor that few seem to accept. If it is true that we really have a duty to intervene and to bear such high risk, then that must be explained on basis of a very strong duty to assist others. And, maybe we have such a strong duty, but that is not a widely held view.

22 "Rather, humanitarian military forces are expected instead to incur some risk to themselves, and surely to avoid even inadvertent commission of the kinds of act they are intervening to prevent" (Lucas 2003, p. 78).

affluent countries—have a duty to compensate those who volunteer to take on the additional risk of intervening. While we have no duty to intervene as such, unless the intervention could be performed with a very moderate level of risk, there could be a duty to finance the intervention in the hope that willing soldiers will come forward ready to bear the necessary risk associated with a successful intervention.[23] And if the intervening soldiers in this way have contracted to take on the additional risk, they would have a contract-based duty to do so.[24] A careful examination of this option would require a paper on its own.[25]

Bibliography

Barry, C. and Øverland, G. (2009), "Responding to Global Poverty: Review Essay of Peter Singer's *The Life You Can Save*," *Journal of Bioethical Inquiry,* 6:2, 239–47.

Hurka, R. (2005), "Proportionality in the Morality of War," *Philosophy & Public Affairs*, 33, 34–66.

International Commission on Intervention and State Sovereignty (ICISS) (2001), *The Responsibility to Protect* (Ottowa: International Development Research Centre).

Lackey, D.P. (2006), "The Good Soldier versus the Good Cop: Counterterrorism as Police Work," *Iyyun: The Jerusalem Philosophical Quarterly*, 55, 66–82.

Lucas, G.R. (2003), "From jus ad bellum to jus ad pacem" in Chatterjee, D.K and Scheid, D.E. (eds), *Ethics and Foreign Intervention* (Cambridge: Cambridge University Press).

Mapel, D.R. (2004), "Innocent Attackers and Rights of Self-Defense," *Ethics and International Affairs*, 18:11, 81–8.

McMahan, J. (1994), "Innocence, Self-Defence and Killing in War," *The Journal of Political Philosophy*, 2, 193–21.

McMahan, J. (2008), "Debate: Justification and Liability in War" *The Journal of Political Philosophy*, 16, 227–44.

---

23   The duty to provide financial support for humanitarian interventions would have to compete with other goals we might have a duty to finance, like assisting the global poor.

24   And if my arguments in this chapter are correct, it does not suffice simply to observe that in some countries, such as the Netherlands, nearly all military activity consists of humanitarian or peacekeeping operations. Although such activities are plausibly part of their job description, and one they are likely to be aware of, it does not follow that they have a higher duty to take risk when engaged in them, as the police at home may not even be required to do that.

25   This paper was first presented in the workshop "New Wars and New Soldiers: Ethical Challenges in the Modern Military" at CAPPE, University of Melbourne in August 2008. I am grateful for comments and suggestions from its participants, and in particular for written comments by Jessica Wolfendale. I would also like to thank Andrew Alexandra and Steven Curry for helpful discussions.

Miller, R.W. (2003), "Respectable Oppressors, Hypocritical Liberators: Morality, Intervention, and Reality," in Chatterjee, D.K. and Scheid, D.E. (eds), *Ethics and Foreign Intervention* (Cambridge: Cambridge University Press).

Norman, R. (1995), *Ethics, Killing and War* (Cambridge: Cambridge University Press).

Øverland, G. (2005a), "Killing Civilians," *European Journal of Philosophy*, 13, 345–63.

Øverland, G. (2005b), "Contractual Killing," *Ethics*, 11:5, 692–20.

Øverland, G. (2006), "Killing Soldiers," *Ethics & International Affairs*, 20:4, 455–75.

Singer, P. (1972), "Famine, Affluence and Morality," *Philosophy and Public Affairs*, 1, 229–43.

Singer, P. (2009), *The Life You Can Save* (Melbourne: Text Publishing).

Steinhoff, U. (2008), "Debate: Jeff McMahan on the Moral Inequality of Combatants," *The Journal of Political Philosophy*, 16, 220–26.

Tripodi, P. (2006), "Peacekeepers, Moral Autonomy and the Use of Force," *Journal of Military Ethics*, 5, 214–32.

Walzer, M. (2006), *Just and Unjust Wars: A Moral Argument with Historical Illustrations*, 4th edn (New York: Basic Books).

Williams, B. (1998), "Consequentialism and Integrity," in Samuel Scheffler (ed.), *Consequentialism and its Critics* (Oxford: Oxford University Press).

Williams, B. and Smart, J.J.C. (1973), *Utilitarianism, For and Against* (Cambridge: Cambridge University Press).

Chapter 5

On a Duty of Humanitarian Intervention

David Lefkowitz

Perhaps the most discussed topic amongst just war theorists during the 1990s was the moral (and legal) justifiability of armed humanitarian interventions. Not surprisingly, that changed after the 9/11 terrorists attacks and the invasions of Afghanistan and Iraq, with topics such as the morality of terrorism, torture, and preventive war receiving the lion's share of attention. Nevertheless, for reasons both good, such as the International Commission on Intervention and State Sovereignty's endorsement of a limited duty of intervention in its report, *The Responsibility to Protect*, and bad, such as the conflict in Darfur, the morality of humanitarian intervention remains a live topic amongst theorists and practitioners alike. A striking feature of the contemporary discussion is the extent to which one prominent feature of the debate during the last decade of the twentieth century, namely the tension between intervention and respect for sovereignty, is no longer at issue. Theorists writing today almost universally endorse the moral permissibility of humanitarian intervention, at least under certain conditions. Disputes remain, of course, as to precisely what those conditions are, who enjoys the right to carry out a humanitarian intervention, and exactly how to balance those moral considerations that count in favor of it, e.g. protecting those subject to a genocidal campaign, against those that count against, e.g. preserving some degree of respect for the rule of international law and state sovereignty. Yet the fact that hardly any theorist, and not a few practitioners, would deny that under certain conditions humanitarian intervention is morally permissible, and indeed, something that certain agents have a right to do, marks a significant and, in my view, positive change from the previously dominant position.

With a limited moral right to humanitarian intervention now relatively uncontroversial, a growing number of just war theorists have begun to add their voices to those of the few thinkers who previously argued for a duty, and not just a right, to wage such a war. For the most part, these theorists focus on demonstrating that the same moral argument that establishes a right to carry out a humanitarian intervention also entails a duty to do so (Bagnoli, 2006; Buchanan, 1999; Davidovic, 2008; Lango, 2001; Tan, 2006). Unlike in the case of a right, however, an agent subject to a duty must bear certain restrictions on his freedom that he would not have to bear were he not subject to that duty. In contrast, though an agent's exercise of a right may entail the acquisition of certain restrictions on his or her freedom, since the agent is free not to exercise that right the possession of it does not necessarily entail any costs for the agent. This asymmetry between rights and duties, together with the assumption (shared by many theorists) that restrictions on individual freedom stand in need of justification, suggests that the defense of a duty of humanitarian intervention calls for an additional step not required for the establishment of a right to humanitarian intervention, namely a demonstration that the conduct of such a war will not impose excessive or unreasonable costs on those who undertake it.

The purpose of this paper is to offer just such a demonstration. I begin with a critical appraisal of separate attempts by James Pattison and Jovana Davidovic to show that in any case where an armed humanitarian intervention meets the jus ad bellum criteria for a right to wage war, the cost of carrying out that intervention will not be unreasonable. I then consider Cécile Fabre's argument that the cost to those who participate in a humanitarian intervention almost always exceeds what it is reasonable to require of them, with the consequence that such wars are almost always supererogatory, or beyond the call of duty. Drawing on the idea that all persons (at least) have basic moral rights, and that all moral agents have a correlative duty to do their fair share to see to it that all persons securely enjoy these rights, I maintain contra Fabre that neither the cost to individuals of acquiring the skills necessary to carry out an armed humanitarian intervention, nor the risk of suffering harm that individuals bear as a result of participating in such a war, constitute an excessive or unreasonable burden. With respect to the first of these two claims, I contend that agents have no moral grounds for objecting to a *pro tanto* duty to bear their fair share of the burdens involved in applying and enforcing the law, even if that requires several years of service, especially if they stand little risk of serious bodily injury or death as a result. The same conclusion holds, I maintain, in the case of an armed humanitarian intervention. As for the second of the two claims I defend here, the key premise concerns how we ought to conceive of the harms likely to befall some of those who carry out the intervention. I argue that we should do so in terms of the likelihood for any individual with a duty to contribute his or her fair share to ensuring that all enjoy their basic rights that he or she will suffer harm as a result of helping to conduct a humanitarian intervention. So conceived, the likelihood of suffering harm as a result of participating in a humanitarian intervention that meets the jus ad bellum criteria for a just war is quite small. If correct, these arguments entail that the cost to individuals of waging an armed humanitarian intervention are not ones it is unreasonable to require them to bear.

* * *

James Pattison (2008, 272) attempts to rebut the claim that the cost of carrying out a humanitarian intervention almost always renders such conduct supererogatory by arguing that "an intervener would not have a *right* to intervene if intervention is excessively costly to its people." While this is true—if an agent has no right to intervene then, necessarily, he has no duty to do so—it is also beside the point. What must be shown is that where an agent has a right to intervene he also has a duty to do so, and that conclusion does not follow as a matter of deontic logic. Pattison fails to rule out the possibility of cases in which the costs of conducting an intervention are high enough to render such a course of action optional, but not so high that they render such a war unjustifiable. Perhaps it is true, as Pattison (2008, 270) contends, that if carrying out an intervention will leave a state "unable to provide vital services, such as clean water provision, for its home population," then that intervention cannot meet the jus ad bellum criterion of proportionality or, given the likely opposition from the home population, the criterion that the war stands a reasonable chance of success. Nevertheless, it seems not only possible but plausible to think that situations will arise in which intervention would impose a serious burden on a state and its subjects without coming anywhere near making it impossible for that state to provide its subjects with vital services. In other words, it is far from obvious that the category of interventions that excessive cost renders optional, i.e. ones that agents have no duty to conduct, is coextensive with the category of interventions that excessive cost renders impermissible, i.e. ones that agents have no right to conduct. Granted, without a precise formulation of what counts as an unreasonable cost this rejoinder may seem to be little more than an assertion. Still, the fact that many people endorse the existence of supererogatory acts in other contexts suggests that the burden of proof lies with those, like Pattison, who reject the possibility of supererogatory humanitarian interventions.

In her defense of a duty of humanitarian intervention, Jovana Davidovic (2008, 140–1) denies that a humanitarian intervention can meet the jus ad bellum requirement that it stand a reasonable chance of success and yet also impose an unreasonable cost on those who wage it. Specifically, she writes that,

> the reason this concern about danger to the intervener is not mentioned explicitly [among the conditions for a just war] is because the role of the notion of danger to oneself in calculations of minimal decency can be captured by an appeal to the condition of reasonable chance of success. ... [T]o be able to say that there is a reasonable chance of success we have to assume that the intervener can alter the conditions on the ground without thereby losing some unreasonable number of soldiers. In other words, due to the interdependence of danger to self and chance of success it is not possible to say that all the conditions of justice of war have been met, but that the expected loss of life of soldiers of the intervening nation, i.e. the danger to self, could invalidate the obligation to intervene.

I believe this line of argument contains two errors. First, the class of cases in which an agent stands a reasonable chance of successfully intervening to protect others from some harm is not co-extensive with the class of cases in which the cost to the agent of attempting to do so is one it is reasonable to require him to bear. For example, in some cases the odds of an agent successfully rescuing another from an imminent harm may be quite high, but so too may be the odds that in doing so the agent will suffer a harm that will substantially shorten his life, e.g. by exposing him to a great deal of radiation or to certain fatal diseases. Similarly, the bulk of the danger posed by a military mission may arise only after the achievement of its aim, as in the case of planes having to navigate fully-awakened air defense systems after they have completed their bombing runs. Finally, it is conceivable that a war of humanitarian intervention could stand a reasonable chance of success and yet also prove too costly, say if it left the intervening state ill-prepared to defend itself against another looming threat.

A second and more important reason to reject Davidovic's reasoning, however, is her assumption that the harms suffered by an intervening force should be understood solely in terms of the cost they impose on the state. The death of a few soldiers may do little to impair the state's ability to pursue its ends, but it is obviously a total loss for those who are killed, and will likely be thought by many to be an unreasonable cost to require these soldiers to bear. Earlier in her paper, Davidovic argues that state sovereignty is valuable only to the extent that it serves to protect basic human rights. This suggests that Davidovic endorses value individualism, i.e. the view that only the lives of individual human beings have ultimate value, with collective entities such as states valuable to the extent that they contribute to the value of individual's lives. Since an assessment of the costs involved in carrying out a humanitarian intervention solely in terms of the effect they will have on the state runs counter to value individualism, it is surprising to see Davidovic argue as she does. Indeed, even many of those inclined to assign independent value to the flourishing of collective entities are unlikely to go so far as to treat individuals as having value only to the extent that they contribute to the state's flourishing. It follows that for all but the most extreme communitarians the assessment of the burdens involved in conducting a humanitarian intervention must take into account setbacks to the interests of the individual combatants that carry it out, as well as the state in whose military they serve.

John Lango (2001, 186) suggests that doing so leads to a moral paradox of intervention: "even if it is obligatory for (the citizens of) a state (collectively) to intervene, it can still be only supererogatory (individually) for its citizens." The alleged paradox disappears, I believe, if we recognize that even when a group acts collectively the costs of doing so ultimately consist of setbacks to the interests of the individual members.[1] In some cases, collective action allows for, or coincides with, the distribution of costs in such a way that they are not unduly burdensome for any member of the group, though they would be were any one member to bear the costs in full. In other cases this is not so, either because the costs remain excessive even when distributed amongst the group or because it is impossible to distribute the costs. It is easy to see how we can be lead to posit the alleged paradox Lango identifies if we begin by asking whether *states* have a duty to engage in humanitarian intervention. Like Davidovic, Lango notes that the deaths of a few (hundred?) combatants will often have a negligible effect on the state, though unlike Davidovic he also recognizes that the cost to those individual

---

1   This is so even where the direct cost is to the ability of the group to functional well as a group, as in the case of reductions in the efficacy and/or efficiency of state institutions.

combatants are enormous. But if we begin instead by asking whether individuals have a duty to participate in humanitarian interventions via institutions designed to facilitate collective action, including (but not necessarily limited to) states, we do not encounter any paradox. Rather, all of the costs are understood in terms of the individuals who ultimately bear them, in which case Lango ought to conclude that (the citizens of) states sometimes have a duty to shoulder certain economic burdens involved in carrying out a humanitarian intervention (e.g. to provide armored personnel carriers to those prepared to carry out the intervention, as in Lango's own example), but no duty to serve in such a war.[2]

The issue of the costs to the state of carrying out a humanitarian intervention are not uninteresting, even for those who deny that its flourishing has any value independent of the contribution it makes to the lives of individual human beings. I set this issue aside here, however, in order to focus on the costs humanitarian intervention imposes on those individuals who carry it out, and in particular the question of whether these costs are so high as to render such actions supererogatory, rather than obligatory.

* * *

Though she maintains that the duty to rescue encompasses a duty to kill under certain circumstances, Cécile Fabre (2007) argues that it follows from the reasonable cost proviso on the moral duty to assist others in need that no one bears a duty of humanitarian intervention. Given that those conducting a humanitarian intervention often face serious risks of harm, and that no agent "is under a duty to provide assistance to others at the cost of their own prospects for a flourishing life," Fabre (2007, 369) concludes that states may justifiably employ only volunteers to wage a humanitarian war. Of course, it is conceivable that in some cases a state might carry out a humanitarian intervention while exposing its individual combatants to little or no risk; arguably, NATO's intervention in Kosovo, and the United Nation's intervention in East Timor, provide actual examples of such "low-risk" interventions.[3] Yet even in these cases Fabre (2007, 372) argues against a duty of humanitarian intervention. She reasons as follows: humanitarian interventions will be low-risk only if they are conducted by those with specific skills the acquisition of which requires considerable investment; or in other words, only if such interventions are undertaken by military professionals. However, while

> individuals are, under some circumstances, under a duty to provide a service
> to the needy, conscripting the able-bodied into specific professions for the sake
> of the needy would undermine their autonomy and thus their prospects for a

---

2   I challenge this second claim below.

3   I set aside here legitimate concerns that low-risk interventions such as the NATO air campaign against Serbia almost always violate the jus in bello principle of discrimination (or non-combatant immunity) and its legal analogs.

flourishing life. More generally, individuals are not under a duty to acquire the skills with which they might be in a position, at some stage, to help the imperiled—any more than they are under a duty to ensure that they have surplus material resources just in case they might chance upon someone who is starving.

I challenge several premises in Fabre's argument below. First, however, I want to consider the possibility of rejecting it because it leads to an absurd conclusion, namely that conscription is morally unjustifiable even in the case of morally justified wars of national defense.

Wars of national defense often require that those who fight them bear a high risk of death or serious injury, or at least considerable sacrifices of time and effort necessary to acquire those skills that allow them to wage war at relatively little risk to themselves. Fabre claims that these costs render participation in an armed humanitarian intervention supererogatory. If true, then in the absence of any morally relevant difference between wars of humanitarian intervention and wars of national defense, Fabre's argument entails that participation in a war of national defense is also supererogatory. If, as some will likely maintain, such a conclusion is absurd, and again, if there are no morally relevant differences between the two types of war at issue, then this calls into question the truth of Fabre's argument against a duty of humanitarian intervention.

Are there any morally relevant differences between humanitarian intervention and a war of national defense? Someone might argue that a war of national defense is not an instance of assisting the needy, and that therefore the reasonable cost proviso does not apply. I think this claim erroneous. Either a combatant fighting in a just war of national self-defense is attempting to aid his co-nationals by protecting them against unjust aggressors, or he is engaging in personal self-defense, or both. Insofar as the duty to aid the needy is necessarily qualified by the reasonable cost proviso, that is true regardless of whether those who are to be rescued are co-nationals or not. As for the case of personal self-defense, I know of no compelling argument for either a duty of self-preservation or a duty to assert one's moral rights, e.g. to just political institutions, regardless of how much or how little it will cost one to do so.[4]

Of course, there may be grounds other than a duty to provide aid or to rescue those in need that serve to justify conscription for a war of national self-defense, but not for a war of humanitarian intervention. Co-nationals, or fellow citizens, may share a special relationship analogous to that of immediate family members, one they do not share with citizens or subjects of other states. Alternatively, co-nationals may owe one another, but not foreigners, a duty of fair-play or reciprocity. Elsewhere (Lefkowitz, 2006) I criticize both of these positions, but were they to

---

4   In certain circumstances agents may have duties *to others* that require them to pursue self-preservation, but that is not the same as having a duty to oneself to preserve one's life or to live under just political institutions.

be true they might well entail a duty to fight in a war of national defense, or at least set the bar for when the cost of doing so is unreasonable far higher than in the case of a duty to provide aid to any moral agent in need.[5] Moreover, my own view (which Fabre may well share) is that the denial of a duty to participate in a war of national defense in cases where doing so will be quite costly does not constitute an absurd conclusion. I will not belabor the point here, however, since I argue below that in many cases of armed humanitarian intervention the cost to individuals of participating in such a war is not an unreasonable restriction on their individual liberty or their pursuit of a flourishing life.

Though intuitively attractive, I contend that Fabre's argument against a duty of humanitarian intervention is mistaken with respect to both low-risk interventions as well as those likely to result in a moderate number of injuries and deaths to the persons conducting them. Consider, first, Fabre's (2007, 372) argument against a duty to carry out low-risk humanitarian intervention. The key premise in this argument is that "conscripting the able-bodied into specific professions for the sake of the needy would undermine their autonomy and thus their prospects for a flourishing life." I think this claim overly broad and therefore false. Suppose, what I take to be fairly uncontroversial, that people can securely enjoy their basic moral rights only if they live under certain types of institutions; for example, ones charged with making, applying, and enforcing the law. If so, and if peoples' basic moral rights correlate to duties on others to do their fair share to see to it that they securely enjoy those rights, then it follows that all those with the duties in question are morally required to contribute their fair share to the creation and operation of those institutions. This might well require limited service in the police, judiciary, or of special interest here, the armed forces. Keep in mind that we are assuming a case in which humanitarian intervention will impose little cost on those who carry it out *if* they are military professionals. By hypothesis, then, the combatant engaging in a humanitarian intervention faces little more risk to his future flourishing than does the typical judge or traffic enforcement officer in a generally law-abiding and moderately just society. What is at issue is the cost involved in acquiring and employing certain skills over a limited, though not insubstantial, portion of one's life. I see no reason to think that, if it constitutes one's fair share of the collective moral task of providing all with the secure enjoyment of their basic moral rights, a period of service amounting perhaps to several years in total necessarily undermines a person's autonomy or their prospects for a flourishing life. Moreover, I suggest that if the only way to ensure the creation and operation

---

5   Strictly speaking a moral duty to participate in a war of national defense is not the same thing as a duty to obey a law mandating military service in a war of national defense. Nevertheless, the reasons for rejecting the latter (at least for many citizens of modern states) also provide reasons for rejecting the former.

of these morally necessary institutions is through coercion, then there exists a *pro tanto* moral justification for doing so.[6]

Obviously universal service in the judiciary, internal or external security forces, etc, is often unnecessary because there are a sufficient number of agents willing to fill those positions voluntarily. Or rather, they are willing to do so conditional on the receipt of a certain level of compensation, compensation that is provided primarily by those freed from the burden of service by these individuals' willingness to work as judges, police officers, and so on. Why should the same not be true of serving in the state's armed forces? That is, the duty to contribute one's fair share to the collective moral task of seeing to it that all enjoy their basic rights may require from many only that they transfer a portion of their economic resources to pay those willing to serve in the armed forces, at least if such transfers will result in a sufficient number of volunteers. Admittedly, Fabre focuses on cases where only coercion can produce individuals with the necessary skills. But suppose it is only possible to provide a level of compensation sufficient to entice the necessary number of judges and police officers by employing a coercive system of economic transfers; i.e. a tax. I suspect few will find such a system morally problematic when employed for this end. Nor, I submit, should we be troubled by the use of this means to generate an army of volunteers willing to conduct a humanitarian intervention.

In contrast to Fabre I contend that where it is necessary to ensure that people securely enjoy their basic moral rights a system of conscription aimed at ensuring that a sufficient number of individuals acquire those skills necessary to achieve this end does not violate the autonomy of any of those agents. If on the basis of our concern to respect others' autonomy conscription troubles us in a way that taxation does not, that may be due largely to two assumptions, each of which should be called into question. The first of these is the conception of military service as a full-time commitment that requires putting aside almost all of the other projects one might pursue. It is not clear how accurate this depiction of military service is – for example, many of the personnel in the US military serve on a part-time basis – nor, more importantly, whether military service must be a full-time commitment. Consider, too, that many countries grant conscripts a certain degree of flexibility in exactly when they must complete their military service; for instance, permitting them to complete a university degree before discharging their military obligation. The second, and more important, assumption is the belief that the effects on most individuals' ability to pursue a flourishing life that follow from active-duty military

---

6   In my view, perfect duties, by which I mean ones that an agent must discharge on every occasion when she can do so, provide a *pro tanto* justification for others' use of coercion to ensure that the agent does what she has a duty to do. That *pro tanto* justification can be overridden or defeated if, for example, enforcement is likely to produce a morally worse outcome than non-enforcement, or if by its very nature the duty is one that cannot be enforced, as in the case of a duty to apologize, or a duty to display gratitude, both of which require that the agent act with a specific motive.

service far exceed the limitations imposed by the payment of taxes. That this is always or even often so is far from clear to me; moreover, it is important to note that its truth turns on the length of service and the level of taxation, and perhaps other factors as well. Deployment overseas obviously places great burdens on an agent, but then so too do the taxes an agent must pay over the course of his or her lifetime. Military service often does require individuals to bear a risk of harm in a way that taxation does not, of course. But remember, in the case of low-risk humanitarian interventions what is at issue is the cost of acquiring the skills of a military professional, not the likelihood that one will suffer harm while fighting a war.[7]

What of Fabre's (2007, 372) claim that "individuals are not under a duty to acquire the skills with which they might be in a position, at some stage, to help the imperiled." This seems true of paradigmatic cases involving a Samaritan duty. In such cases, the risk of significant harm is immediate and typically abnormal both in terms of the frequency with which people are exposed to it and the number of people who suffer such exposure. It follows from the very nature of the risk that it cannot be addressed by institutional means, and therefore the duty falls on those who happen to confront the person in danger. The foregoing discussion highlights a different type of positive duty, however, one that requires agents to do their fair share to ensure that all are secure from the most grievous harms or wrongs endemic to a state of nature, or from rule by those who manifest no principled commitment to respect for others' basic moral rights. All are at risk of suffering these harms, and perhaps more importantly, the risk can be alleviated to a considerable extent by the creation and operation of various institutions. One question that must be addressed, then, is whether a duty of humanitarian intervention is properly conceived of as a kind of Samaritan duty or as a positive duty correlative to peoples' basic moral rights. Fabre assumes the former in her analysis of a duty of humanitarian intervention. However, many contemporary theorists (Shue, 1980; Orend, 2002) argue that while an individual's first claim to the secure enjoyment of her basic moral rights is against her state or fellow citizens, in the event that this claim goes unmet the individual has a claim against

---

7   Our ability to determine whether a given humanitarian intervention will be a low-risk one may be quite unreliable, of course. If true, that fact may warrant a strong presumption against conducting any such operation. This point is distinct from the one at issue in the text, however, since Fabre maintains that even if we can be certain that a given intervention will be low-risk there is still no duty to carry it out, or more precisely, that no one has a duty to acquire those skills that would make an intervention low-risk, while I argue for the opposite conclusion. Moreover, the level of risk borne by combatants conducting a humanitarian intervention depends a great deal (though certainly not entirely) on choices their leaders make regarding both the specific mission and the type and number of military assets committed to the operation. While caution should be the byword when considering whether to undertake a humanitarian intervention, it would be an exaggeration to claim that we can never judge with sufficient confidence that a particular intervention will be a low-risk one.

the international community, understood to consist of all states, or perhaps more accurately, all moral agents. If these theorists are correct, then it follows that there is a duty to acquire the skills necessary to carry out a humanitarian intervention if this can be done at a reasonable cost, as I maintain that it can.

While I reject the view that all cases that call for armed humanitarian intervention are properly conceived of as instances in which others have only a Samaritan duty to intervene, I do not mean to assert the opposite. Even in a world with a reasonably well functioning institution designed specifically to address the need for armed humanitarian intervention, circumstances might arise in which, say because of its other on-going commitments, this institution cannot respond adequately to a campaign of ethnic cleansing. Here the features of a paradigmatic case involving a Samaritan duty are all present: people face an immediate risk of significant harm that cannot be ameliorated even by well designed institutional mechanisms. Do all agents have a duty to acquire the skills of military professionals just in case such a situation occurs? Here I largely agree with Fabre: given that they have done their fair share to create and maintain the institution charged with responding to the need for humanitarian intervention, agents have only a Samaritan duty to acquire the skills (and other resources) necessary to help the imperiled should the aforementioned institution be unable to do so.[8] To the extent that the acquisition of these skills will be excessively costly, and given that the successful undertaking of a (low-risk) intervention requires possession of these skills, it follows that such agents will have no duty to carry out an intervention in the situation in question. In short, the duty to ensure that all persons securely enjoy their basic moral rights does entail a *pro tanto* duty to acquire those skills necessary to serve in institutions without which it is extremely unlikely that all will securely enjoy their basic moral rights. That includes institutions designed to carry out humanitarian interventions, or at least low-risk ones, for the same reason that agents have a duty to contribute their fair share to the creation and operation of domestic institutions that enact, enforce, and apply law. As long as they fulfill that duty, however, agents have only a Samaritan duty to devote additional time and effort to developing skills or acquiring resources that may be necessary to rescue others from perils that go unaddressed even when these institutions function reasonably well.

* * *

Of course humanitarian intervention often does require those who undertake it to bear significant risks of death or serious injury, at least if they wish to conduct

---

8   I say I largely agree, because Fabre maintains that there is no duty to acquire such skills, while I claim that there is a duty to do so if it can be done at a reasonable cost. In a nearly just world with well-functioning institutions designed (in part) to address situations that call for humanitarian intervention, the likelihood of needing to intervene because those institutions fail to do so will be so small that almost any cost involved in acquiring those skills will be one it is unreasonable to demand that people bear.

their intervention in line with the moral principles of jus in bello and their legal analogs. Does it follow, as Fabre maintains, that given the reasonable cost proviso on the duty to assist others in need there is no duty of humanitarian intervention in such cases? It depends, I suggest, on how we ought to conceive of the risk of harm involved in conducting a humanitarian intervention. Fabre (2007, 370) writes, rightly in my view, that "it would not be plausible to hold that individuals cannot be held under a duty to incur *any* life-threatening risk, however minute, for the sake of another." What matters when it comes to calculating whether the cost of providing aid to others is reasonable is not only the potential harm to which an agent will be exposed but also the likelihood that he or she will actually suffer that harm. That likelihood is the sum of a large number of variables, of course, but the one I want to focus on here concerns the pool of agents with the potential to provide assistance, and in particular, to participate in a humanitarian intervention. Specifically, I will argue that if the pool of those able to contribute to such a war is large enough, then the likelihood of any one of them dying or suffering a grievous injury while fighting in it is low enough that it is reasonable to demand that they shoulder that risk.

Earlier I proposed that all agents have a duty to contribute their fair share to the collective moral task of ensuring that all people securely enjoy their basic moral rights. If correct, this claim entails that the pool of potential participants in a humanitarian intervention is coextensive with the class of all moral agents.[9] Fairness requires that the costs involved in conducting a humanitarian intervention, and in particular the risk of suffering some serious harm, be distributed equally among those in this pool. Or, more realistically, the risk of harm should be distributed as equally as possible. On the assumption that training and deploying all of the agents in the pool of potential participants is neither feasible nor morally desirable, I suggest that the use of a lottery to determine who should actually serve in a humanitarian intervention provides the best account of how the duty in question ought to be discharged.[10] If, as will sometimes be the case, only a small percentage of those in the pool of potential participants in a humanitarian intervention will actually need to do so in order for that intervention to be successful, and if only a small percentage of those will actually suffer serious harm as a result of their participation, then for each agent in the pool the risk of

---

9   It may be that there are some moral agents with physical or mental disabilities that render them practically unable to participate in a humanitarian intervention, or any other type of war. Agents that are freed from an obligation to participate in a humanitarian intervention on these grounds should not be confused with those that have the duty in question, but that are also subject to other moral duties that sometimes outweigh or defeat their duty to bear their fair share of the burden involved in conducting a humanitarian intervention.

10   For a defense of lotteries as a just means for distributing benefits and burdens among those with a claim or liability to them, especially in cases where the benefits or burdens cannot be distributed, see Stone, 2007.

harm posed by a humanitarian intervention will be quite small. In other words, for each agent with a duty to participate in a humanitarian intervention, the cost of doing so—understood in terms of the likelihood of suffering grievous harm and the other opportunities forgone while engaged in military service—is not an unreasonable one. Therefore the reasonable cost proviso on a duty to take positive steps to ensure that others securely enjoy their basic moral rights does not entail that there is no duty to participate in a humanitarian intervention. And while more needs to be said to establish this conclusion, I believe it also follows from this line of argument that a state, or at least a legitimate one, has a right to use a fair method of conscription to staff a military force capable of carrying out moderately dangerous humanitarian interventions.[11]

The practice of mandatory immunizations, which most people appear to view as morally unproblematic, provides some intuitive support for these conclusions. In the case of many immunizations there is a small probability of serious harm to the person who receives the immunization. Nevertheless, it might be that one has a duty to do one's fair share to protect others from the serious harm that will occur if no one, or indeed, not enough people submit to an immunization. Likewise, in at least some cases of humanitarian intervention there is a large population at risk of suffering severe harm or death, and a means for preventing this outcome that will result in severe harm or death to only a few of those in a position to contribute to its prevention. If we think that people have a duty to submit to immunizations, or at least that the state has a right to command people to submit to immunizations, and to compel them to submit if they refuse, then I contend that we ought to draw the same conclusion when it comes to these cases of humanitarian intervention.

Someone might challenge this argument by analogy by pointing out that the person who submits to immunization also stands to benefit from it, which is far less likely to be true of those who carry out a humanitarian intervention. Yet it seems to me that even someone who is only a carrier of a disease still has a duty to submit to immunization, since his failure to do so poses a risk to others even if it poses no risk to himself. Similarly in the case of humanitarian intervention, it is the benefit to others that generates the duty, and any benefits to oneself that follow from discharging it are a happy addition, not a condition for the being under the duty in the first place. Alternatively, it might be said that the carrier of

---

11   There is some reason to believe that the larger the force carrying out the intervention, the smaller the probability that any one of them will suffer harm. This is so not only because those opposed to the intervention will have more targets from which to choose, thereby reducing the probability that any particular individual will be a target, but also because a larger force will do a better job at disrupting and deterring those who might want to attack them. If this is right, then there may be a tradeoff between reducing the likelihood that those participating in a humanitarian intervention will suffer grievous harm or death as a result, and reducing the number of agents with a duty to acquire the skill set of a military professional (at a not insignificant cost to their ability to pursue other ends).

a disease poses a threat to others, while the refusal to intervene does not, and that this explains why a carrier has a duty to submit to immunization while no one has a duty to participate in a moderately dangerous humanitarian intervention. We must be careful not to be misled by a metaphor here: the carrier of a disease does not literally threaten others, since the threat he poses is not the result of any intentional action on his part. But just as a person may act wrongly if he knows he is a carrier and refuses to submit to immunization, so too agents in a position to carry out a humanitarian intervention at a reasonable cost may act wrongly if they fail to do so. In both cases the agents in question allow others to suffer a harm that they could prevent (with cooperation from a sufficient number of others) without exposing themselves to much risk of harm.

It may appear that one morally relevant disanalogy between immunization and humanitarian intervention is the certainty of achieving a good outcome. States typically, and probably rightly, mandate immunizations only in those cases where they have a justifiably high level of confidence that this will protect their populations from the targeted disease. A similarly high level of confidence in the likelihood that intervention will successfully prevent or end genocide or ethnic cleansing will frequently be unjustifiable. In some cases a lesser degree of confidence may reflect the fact that humanitarian intervention would be morally problematic, since likelihood of success plays a crucial role in the moral justification of any war. Nevertheless, this apparent disanalogy does not undermine the defense of a duty to carry out humanitarian interventions offered here. Rather, it serves only to highlight a condition on that duty, one that holds with respect to many other moral duties including a duty to submit to immunization, namely that it obtains only in cases where the likelihood of achieving a morally desirable outcome exceeds a certain threshold.

An agent's duty to participate in a humanitarian intervention may also be conditional on his ability to discharge it without compromising certain special obligations he has to his compatriots. For example, combatants may have no obligation to carry out an armed humanitarian intervention—indeed, they may have a duty not to do so—if it will expose their compatriots to a significant risk of armed invasion, or leave their state with few resources with which to respond quickly to a major natural disaster. Even if this is the case, the existence of a special obligation to compatriots does not undermine the argument I offer here for a duty of humanitarian intervention. Nor, I suggest, is there reason to believe that in practice special obligations to compatriots will always override the duty of humanitarian intervention. After all, few if any of the armed interventions conducted over the past two decades that can plausibly claim to have been humanitarian ones (a contested category, I admit) have exposed citizens of the intervening states to the risks identified above. Moreover, it may be that the duty to contribute one's fair share to ensuring that all enjoy their basic moral rights requires citizens of many states to contribute more to facilitating armed humanitarian interventions than they currently do. Among other things, this might require an increase in military spending so that there will be fewer occasions on which combatants' special obligations to their compatriots trump the duty of humanitarian intervention.[12]

If correct, the arguments presented here provide some reason to think that a reasonable cost proviso on the duty to provide assistance to those at great risk of serious harm does not in principle preclude a duty of humanitarian intervention. At present, of course, no procedure exists for selecting participants for armed humanitarian interventions amongst all those with a duty to see to it that others securely enjoy their basic moral rights. The claims defended in this paper provide a justification for creating such procedures, however. Or, seen from a different perspective, they rebut the position that no such procedure need be instituted because, given its costs, humanitarian intervention is never morally obligatory.

Furthermore, even in the absence of a morally adequate procedure for distributing the burdens involved in conducting a humanitarian intervention amongst all those with a duty to contribute, it may still be possible to employ the argument presented here to sustain the claim that individuals, and the states in whose military they (can) serve, have a duty to carry out a particular humanitarian intervention. Specifically, in any case where it is reasonable to believe that humanitarian intervention will pose an equal or lesser risk of harm than what an agent would bear if all moral agents were to actually "place their names in the hat" of the ideal procedure, the cost to that individual of carrying out the intervention is not unreasonable. Or at least this is so if I am right when I claim that, distributed across a very large number of agents, the costs involved in intervention are very low. *Ceteris paribus*, it follows that this agent has a duty to participate in the intervention. So for example, the argument presented here might be used to demonstrate that the British combatants who carried out the May 2000 intervention in Sierra Leone had a duty to do so, even in the absence of a fair process for selecting them for this task from amongst all those with agents with an equal duty to contribute to it.

All else may not be equal, of course. In particular, it may be that the agent in question has already contributed his fair share to the collective task of protecting all persons from genocide, ethnic cleansing, and other types of threats that clearly provide grounds for humanitarian intervention, say by participating in some earlier war of this type.[13] This may even have been true for the British troops that intervened in Sierra Leone. As these remarks indicate, the idea that the risk of harm involved in carrying out a humanitarian intervention should be distributed fairly amongst all those with a duty to contribute to it functions in two distinct

---

12   Michael Gross (2010, 227–8) discusses this point in the context of the European Union's plans for a rapid response force designed specifically to conduct humanitarian interventions, and suggests that the duty to distribute fairly the risks involved in carrying out an armed humanitarian intervention may justify conscription.

13   Some (Murphy, 2002), argue that no duty obtains in such circumstances, while others (Singer, 2004) claim intention is still morally required though the intervening agents also have a claim to compensation from those whose failure to contribute entails a need for the intervening agents to do more than their fair share.

ways in the argument for a duty of humanitarian intervention. First, it plays a key role in establishing that when properly conceived this duty imposes little cost on those agents with the duty. Second, it provides a key premise in an argument for the creation of a specific type of institution that will identify who in particular ought to carry out a given humanitarian intervention.

Bibliography

Bagnoli, C. (2006), "Humanitarian Intervention as a Perfect Duty: A Kantian Argument," in Terry Nardin and Melissa S. Williams (eds), *Humanitarian Intervention* (New York: New York University Press).

Buchanan, A. (1999), "The Internal Legitimacy of Humanitarian Intervention," *Journal of Political Philosophy*, 7:1, 71–87.

Davidovic, J. (2008), "Are Humanitarian Military Interventions Obligatory?," *Journal of Applied Philosophy*, 25:2, 134–44.

Fabre, C. (2007), "Mandatory Rescue Killings," *The Journal of Political Philosophy*, 14:4, 363–84.

Gross, M. (2010), *Moral Dilemmas of Modern War* (Cambridge: Cambridge University Press).

Lango, J. (2001), "Is Armed Humanitarian Intervention to Stop Mass Killing Morally Obligatory?," *Public Affairs Quarterly*, 15:3, 173–89.

Lefkowitz, D. (2006), "The Duty to Obey the Law," *Philosophy Compass*, 1:1, 1–28.

Murphy, L. (2002), *Moral Demands in Non-Ideal Theory* (Oxford: Oxford University Press).

Orend, B. (2002), *Human Rights: Concept and Content* (Peterborough, Ontario: Broadview Press).

Pattison, J. (2008), "Whose Responsibility to Protect? The Duties of Humanitarian Intervention," *Journal of Military Ethics*, 7.4, 262 83.

Singer, P. (2004), *One World* (New Haven: Yale University Press).

Shue, H. (1980), *Basic Rights* (Princeton: Princeton University Press).

Stone, P. (2007), "Why Lotteries are Just," *The Journal of Political Philosophy*, 15:3, 276–95.

Tan, K. (2006), "The Duty to Protect," in Terry Nardin and Melissa S. Williams (eds), *Humanitarian Intervention* (New York: New York University Press).

PART III
New Technologies in the Battlefield

Chapter 6

Emerging Military Technologies: A Case Study in Neurowarfare

Nicholas Evans[1]

There has been considerable recent debate regarding the status of novel military technologies and how the use of these technologies meets the standards of the Just War tradition. These issues are especially problematic within the context of asymmetric warfare, warfare in which the relative military strength opposing forces is significantly different. This paper is an exploration of a particular type of novel military technology, in which soldiers can control military robots or vehicles via a direct neural connection. This emerging field has been labelled neurowarfare (White, 2008, 177), and forms part of a larger discussion regarding how technological advances in the military will shape the ways in which wars are conducted.

It has been argued, for example, that the use of autonomous military robots (AMRs) cannot be justified in war. It is unclear who should be held responsible for the deaths caused by the robots, with responsibility for deaths caused being an important factor in the ethics of war (Sparrow, 2008). Further, the use of Unmanned Aerial Vehicles (UAVs) has come under scrutiny as morally problematic. While UAVs do not suffer from all of the same problems as AMRs—the presence of an operator creates a plausible chain of responsibility—it does share some features with AMRs, such as lack of meaningful opponent to strike back against. A more pressing problem, perhaps, is that attacks via UAV have been condemned not as military actions, but as "targeted killings," a practice condemned almost universally (Downes, 2004).

It has been argued that remote weapons as ethically impermissible because there is no legitimate recourse for less technologically advanced opponents (Killmister, 2008). This would either render the use of remote weaponry impermissible in war, or call into question the ultimate justificatory power of just war theory.

---

1   This chapter started life as a conference paper at the 2008 meeting of the Australasian Association of Philosophy, and was presented in revised form at the "New Wars, New Warriors" symposium at the University of Melbourne in 2008. The author would like to thank Michael Selgelid, Larry May, Christian Barry, Jeremy Shearmur, and Adam Henschke, Jessica Wolfendale, Paolo Tripodi and an anonymous reviewer for their helpful comments on previous drafts of this paper.

This paper will serve as an addition to this emerging body of work by addressing issues raised by another emerging military technology which may solve some of these problems but create more. I will not be dealing with the ultimate impermissibility of remote technologies in war at this time. In some cases, military technologies may leave no recourse to enemies on a grand scale. However, as I will show, we should not underestimate the ingenuity of combatants put into impossible decisions—a core part of asymmetric warfare is that it breeds unconventional tactics. Rather, I will examine the morality of using this new technology from the perspective of the user.

Novel military technologies as the debate so far point at what Khan calls the "Paradox of Riskless Warfare." Khan argues that as asymmetry between opponents grows war becomes more like policing, and thus falls into conflict with the current laws of war (Khan, 2002). However, Khan's concerns rely on a number of as yet unmet conditions. While there is certainly a move towards riskless warfare, there are still evidently many, many risks that US and allied combatants face far beyond the realm of conventional policing, as the tactics their opponents centre on conducting protracted insurgencies over "conventional" war. There is concern, however, that advances in miliary technology will bring this about sooner rather than later (e.g., Sparrow 2009).

This paper will review the current trends in military technology towards automation and long-range, and provide an example of a technology which may, in principle, render warfare riskless. This technology, named a Human Assisted Neural Device (HAND) would allow soldiers to control vehicles and other robotic weapons platforms via a neural connection—that is, they would control robots with their minds. In doing this, I will argue, they escape the problems associated with AMRs, and also the moral and practical limitations of the UAV. But in doing so, they may create an asymmetry so deep that, following Khan's argument, war ceases to be war and rather becomes a policing action. I will then discuss some of the moral, legal and political implications of this.

HANDs

It may seem that there is nothing special about a HAND. After all, we already live in a world where UAVs patrol the sky, the navy tests rail guns (firing depleted uranium via electromagnetic "rails" to achieve energies equivalent to conventional explosives), and the Future Force Warrior demonstration promises technology reminiscent of science fiction. It would be correct to say that considering the mechanization of warfare and the removal of troops from combat and substitution with automatic drones in the case of the air, the next logical step would be to remove ground troops and substitute them with robots controlled directly via the troops from a safe distance.

The literature is already heavily populated by discussions of developments in AMRs, which are still under developments; and Unmanned Robotic Vehicle

(URVs), which are presently seen in popular media through the well known Predator Drone.[2] Ronald Arkin has argued that AMRs are, if programmed sufficiently well, capable of making better decisions than humans by virtue of being less prone to the traditional human errors of judgment that occur in combat: emotional trauma, lack of processing power in stressful situations, hatred, and so on (Arkin, 2007).

However, much of this has been rejected on the basis that AMRs cannot be used in principle because of considerations of responsibility for killings in war. The programmer cannot be held responsible because autonomous weapons systems are a) acknowledged not to be perfect decision makers, and b) will be capable of making choices in a fashion that does not strictly reflect their designer's intentions. The commander who issues orders to the robot cannot be held responsible for the robot, because as robots become more autonomous and is capable of choosing its targets than it becomes unfair for holding a Commanding Officer responsible for those decisions. The robot itself is obviously not a candidate for such responsibility, not being an agent—at this time—in the relevant sense (Sparrow, 2008).

URVs may solve this problem, but create their own practical problems. URVs that rely on manual control or manual control paired with computer assistance will be limited in their applicability through the limited information that can be relayed to a pilot. While this is not necessarily problematic in airborne conflict involving high-altitude bombing, in more complex and varied environments lack of fine control becomes a fatal problem for URVs. Using menu options to perform an action which in a conventional person-in-a-vehicle would require a simple use of a steering column or flight stick can be complicated (Tyabji, 2007).

Moreover, present limitations with bandwidth and signal dropout are being addressed aggressively, but no amount of increase in signal strength and reliability alone can solve the problem of lack of tactile sensation within a cockpit, a vital element in proficient combat flying (Tyabji, 2007). While Arkin and his contemporaries argue that machines make better decisions than humans under fire, if we must use humans than the traits that allow us to fight effectively are vital. Until such factors can be accommodated, it seems that URVs may play a supporting role in combat activities, but will never supplant the need for ground troops.

Neurowarfare is the guidance of robotic military hardware through direct neural connection. Pioneered by the Defence Advanced Research Projects Agency (DARPA) over the last decade, this technology has resulted from new methods of translating the electrochemical processes within the brain into corresponding electrical signals that are able to be interpreted by digital technology. This research has culminated in the creation of a Human Assisted Neural Device (HAND).

The HAND, while technically advanced, is a conceptually simple device. A subject's brainwaves are measured and recorded while they perform certain tasks to map electrical activity in their brain. Then, a Brain–Computer Interface (BCI) is

---

2 I use the term "URVs" because it is not UAVs which I will deal with, but rather unmanned vehicles used in ground, air, or sea warfare.

implanted into the subject's brain, and connected to detect brain activity and turn specified signals from the brain into commands for a machine (e.g., Santhanam et. al., 2006). The pioneering test subject for this technology, an Owl Monkey by the name of "Belle," had her brain waves measured over repeat performances of moving a cursor to move a robotic arm to certain locations to grasp a reward (Garreau, 2006). Once mapped, the BCI in Belle enabled her to use a robotic arm in the same fashion as the cursor.

It should be no surprise that this will inevitably lead to advances which allow for two-way communication between the brain and a linked robot. Cochlear implants are now commonplace technologies, and we can expect the bionic eye in the near future (Lauder, 2010).

Using this technology, robots would be able to be controlled via thought, rather than through cumbersome or inappropriate input devices. A particularly exciting promise for this technology is that it will allow individuals with extreme reduction in motor functioning to interact with the world in new ways. Clinical trials are already in progress for this particular technology on patients with tetraplegia at companies such as Cyberkinetics.

The need for a device which allows for the potential that operators exist as their vehicles rather than work through their vehicles has been recognized as an important component in military operations since the 1960s. The former director of DARPA, J.C.R. Linklider, noted that until such time as genuine artificial intelligence could control weaponry man would need to be "in the loop" of decision making in combat. The symbiosis of humans and computers represents the unification of the massively serial processing of computers with the massively parallel computational power of human minds, and allowed for decisions that no human could make alone (Licklider, 1960).

The HAND, then, is not sui generis but rather the next step in development of this vein of emerging military technologies. That is, the HAND circumvents the problems of AMRs by keeping a "human in the loop" of control and thus someone in charge of the decisions to act—and kill—in war. This allows for the use of robots in war, as while robots can perform all of the functions computers excel at—data collection and representation, communication, surveillance, and automated functions such as defensive manoeuvring where humans may freeze or act badly. It also allows for the humans to do what—for now—humans do best, that is, think tactically, laterally, control multiple systems independently and in parallel, and function with an agency that implies responsibility for actions.

HANDs present all the advantages of URVs in allowing an individual to control a robot in the field, but without the problems of interface management, lack of proprioception and tactile senses, and so on. It is likely that through HANDs, technologies allowing for rapid changes from one URV to another, as well as programming URVs for swarm behaviour (allowing multiple vehicles to act as a single entity much like swarms of insects do) (Scheutz et al., 2005; Scerri et al., 2008), HAND operators will not only be able to support forces in the future, but will supplant these forces.

For now, this is still highly speculative, and relies on advances in information transmission over wired and wireless networks, development of robust wireless communication with improved signal stability and security, and so on. However, these improvements are not a matter of if, but when. In a non-military context, improving the strength and reliability of transmission is a lucrative pursuit in its own right, and with defence applications these advances are likely to happen sooner rather than later (e.g., Etoh and Yoshimura, 2005). There may come a day when HANDs are ubiquitous amongst the powerful militaries of our world.

Neurowarriors and the Permissibility of Killing

The main issue I wish to take up regarding this technology is the relationship between combatants who utilize HANDs as their primary way of waging war (hereafter referred to as "neurowarriors" or "operators") and their less technologically advanced opponents.[3] My primary thesis is that, echoing Khan's predictions, HANDs could create conditions of riskless warfare as they become ubiquitous in modern military operations. I will show this loss of risk changes the moral permission for neurowarriors to kill their enemies.

We might think that the moral permission to kill is central to the ethics of war. The moral permissibility of killing and its limits grounded in necessity, proportionality, and non-combatant immunity, receive a variety of justifications. I do not wish to show that any justification is better than the other—rather, I will show that by any of the major accounts of permission to kill in war, the use of the HAND either does not meet the standards of justification, or does so in an exceedingly limited fashion.

I will view two potential justifications for killing in war, and for each demonstrate why the use of HANDs by soldiers makes killing enemy combatants in asymmetric conflict impermissible. First, I will examine the account in which combatants may kill one another by virtue of being placed in the hostile conditions of war (Walzer, 2000). These claims may be seen to be arguments about the necessity of causing death in war as a method of preserving one's own life and completing the necessary objectives of a war. I will argue that in using a HAND, the situations in which a HAND operator is justified in killing an opponent are significantly limited.

---

3   I will take it as axiomatic that HANDs are unlikely to be used in symmetric conflict at this time, and in cases of asymmetric conflict HANDs are only likely to be used by the more powerful party. This is based on two assumptions. The first is that the major state powers of our time are unlikely to engage in physical conflict with each other over diplomatic or economic manoeuvring, or cyberwarfare. In cases in which conflict occurs between a powerful and a weak state, the weak state is unlikely (at least for very long) to have the communicative infrastructure to be able to sustain operations using HANDs.

Second, I will examine the justification for killing in war, based on the institutional scale of war and the permissibility of killing as part of a larger act of aggression (e.g., Rodin, 2002). I will argue that in confronting a HAND, a large number of combatants can be convincingly seen to be no longer part of the larger act of war, as they are not able to attack or defend objectives in a meaningful way.

Killing in War: The Limits of Necessity and Proportionality

The condition of necessity states that killing in war is permitted as long as such killing is a necessary part of winning a military campaign. That is, one is permitted to kill only those it is necessary to kill in order to prosecute the outcome of a war (Walzer, 2000, 144–5). Proportionality states that even when an operation necessitates killing, killing is only permissible if the goals of war outweigh the cost of the deaths in operations; proportionality is a consequentialist calculation about the costs of killing (Walzer, 2000, 129–30). What I wish to explore is how the HAND, as I have described it, may change the bounds of these conditions to narrow the scope of what counts as permissible acts of killing.

If a soldier receives orders from her officers regarding a particular manoeuvre or operation, killing of enemy combatants we might think is a given. The requirement of necessity states that killing is permissible insofar as it is necessary to the achievement of military goals. Correct strategy will always be part of military operations, as bad strategy can cause even the strongest military body to fail in its objectives. Necessity at the stage of "embodied combatants"—that is, combatants whose flesh and blood bodies are present in the conflict—only requires that we do not shoot the wounded, the innocent bystander, or a combatant who otherwise poses no threat or whose death would be for no purpose (Hurka, 2008).

However, it is not clear that the deaths of soldiers are necessary to the achievements of military operations. The HAND brings the question of the necessity of killing enemy soldiers to the fore. This is surely a question that precedes the HAND, as aerial bombing or basic artillery may seem to carry with it the same problems. However, it seems that once a vehicle is guided via a HAND, and the presence of ground troops begin to disappear, the necessity of killing in war seems evaporates almost completely. There is little to genuinely threaten a HAND, in the sense that its operator may be killed. Moreover, as the HAND becomes ubiquitous, it does not even act in defense of soldiers who may die.

We might, however, think that our permission to kill does not rest on this local level, in which we may only permissibly kill if otherwise we would be killed by the enemy engaged with us in this combat. We might instead think that we are permitted to kill any enemy engaged in the pursuit of the war in question (Rodin, 2002).[4] Another way of putting this is that necessity is not a positive account, that

---

4   I am indebted to Jessica Wolfendale for this objection to a previous draft of this paper.

is, which combatants we can kill; but rather a negative account, outlining those we cannot kill. Anyone who counts as a combatant in the war is fair game, within the bounds of proportionality.

However, this I believe runs into complications when we consider who counts as a combatant. Certainly the civilian and the wounded soldier may not or may no longer be combatants. However, the individual we have sprung upon with their guard down and their gun leaning against a wall is just as defenceless as anyone. The enemy, now with their hands in the air, has no recourse: they have effectively left the status of "combatant." It is true that if we had seen them from afar and had no realistic way to achieve our objectives without killing them, or became engaged in combat with them after they noticed us, we would be justified in killing them. But with their hands in the air, they certainly may have been part of the institutional process of the war, but now they seem to be much more like a noncombatant, helpless as they are. Now, we might think that we have kill them to prevent them, say, from raising the alarm. But unconsciousness is just as effective here as death, assuming our mission is not one requiring a long time to complete, and we cannot otherwise restrain the combatant.

Suzy Killmister argues that AMRs for similar reasons are impermissible from because there is no recourse to the deadly force they present (Killmister, 2008). However, we might think that even if we were to have a justification for using devices such of the HAND from the outset, once we are in war we still have no justification for using them in a lethal capacity. Rather than no recourse to deadly force being a problem, the lack of ability to respond in a meaningful way removes the enemy soldier from their status as combatant. We still might classify them as a threat, and justifiably restrain them, but not as a combatant worthy of killing. They simply cannot take part in the aggression that is being perpetuated.

Who Is Left? Using HANDs and the Legitimate Targets of Military Action

There may be a small number of combatants left who are genuinely able to threaten a force of HANDs, or otherwise retain their status of combatants in the aggression taking place. They are those who threaten the operators of the HANDs; in aid of any ground personnel who are still present in a conflict; those enemy who may be capable of using weapons of mass destruction; those enemy targeting civilians or committing war crimes, or crimes against humanity; and those who are so suitably entrenched that the legitimate aims of the military conflict are unable to be resolved without causing death. I will engage each of these in turn.

There may be enemies who are capable, either by being in the site of a warzone where the neurowarriors themselves are located, or if located abroad, those who function on behalf of a military force to kill the physical operators of the HANDs. Alternately, there may be a case in which a powerful state that possesses HANDs is threatened in a defensive war, and thus HANDs would be permitted to act in the defense of a nation and kill enemy combatants. In this case, those individuals would retain their status of combatants by virtue of having recourse to violence that can legitimately influence the outcome of a war. As such, HAND operators would be permitted to engage in lethal combat either as a matter of self-defense, in aid of their fellow neurowarriors, or otherwise as part of the continuing response to aggression.

It is unlikely in the extreme that our embodied military will disappear overnight, and it may be that they never disappear. In this case, there is cause for HAND operators to act in defense of their fellow soldiers. With the presence of flesh and blood troops on the ground, there will still be combatants who are participating in the practice of war to shoot at. Thus, from the first justification for killing in combat, we may act in defense of others. Moreover, if HANDs do not comprise the substantive part of the force in question, than enemies may still perpetrate aggression against our forces. However, if HANDs are found to be successful and practical in the way they are used, these ground soldiers will become rarer. As they do, these instances will become likewise rarer.

Those who control weapons of mass destruction, we might think, are likewise acceptable targets for neurowarriors.The sheer lethality of weapons of mass destruction, short of being targeted in an area in which the only "casualties" are the robots the neurowarrior control, would be such that neurowarriors would be permitted to act in the defense of others. This may include killing, due to the stakes of such an operation, others in the pursuit of someone with control over weapons of mass destruction. Moreover, the destruction of HANDs on such a large scale may make the legitimate military objectives of a campaign impossible, and thus permit the use of lethal force, as I will discuss shortly.

We might, when we consider humanitarian interventions, think that the use of lethal force is justified in the pursuit of humanitarian goals. In humanitarian interventions, as such, it may be permissible to kill using HANDs. This might be seen as a special case of necessity, in which the objectives of the military action rely on the rapid neutralization of targets to prevent mass killing, or so on. In such conditions, this is limited to proportionality conditions—we might think that killing 1,000 soldiers to save 3 noncombatants problematic. Moreover, there may be other situations, particularly if a HAND were to be destroyed and then more noncombatants killed, that the survival of the HAND is for sufficiently good reason—saving lives—that lethal force is authorized (e.g., McMahan, 2009).

Finally, in the pursuit of a legitimate war, there may be cases where combatants retain their status as combatants in the institutional sense, even though they do not pose a threat to a force. Perhaps there is violent resistance to a just military force, but a few well equipped enemies have set up an electromagnetic bomb which would disable any intervening vehicles. Embodied ground forces are not available as an appropriate intervention owing to urban fighting and serious losses, but a lethal strike at distance would be appropriate. In this case, there does not seem to be any problem. Although the HANDs can operate from a safe distance from the electromagnetic bomb, they may not be able to do so with a nonlethal or nonviolent capacity. As such, they may be permitted to bomb the building in question, as would be the case in conventional war sans neurowarriors.

War in a World of HANDs

What conclusions might we draw from this? The first two mirror Killmister's conclusions regarding robotic weapons (Killmister, 2008). Either these devices force enemies into unconditional surrender, or they force the enemies of a state equipped with such technology to target civilians. But these are not particularly new conclusions—in fact, these conclusions may be seen in current asymmetric wars, though in a lesser capacity. These are both unpalatable options, and may force us to reconsider whether adopting such technology is just. However, we might wonder if such an analysis really pays attention to the resourcefulness of humans when the odds are stacked against them. Though asymmetric warfare in its paradigmatic examples—take guerilla warfare or organized insurgency, for example—is considered abhorrent, we might think that this is the legitimate response to a threat. Merely because such tactics are not the tactics one could use against neurowarriors does not mean that there are not tactics that could work against such a force.

Such technology may institute an "asymmetric" arms race. As HANDs become prevalent in highly funded, developed militaries in countries with the communications infrastructure to support them, countermeasures to these devices will be developed rapidly. Weapons such as electromagnetic pulse (EMP) weapons may become readily available. One potential result of this demand for EMP weapons is the furthering of nuclear proliferation. Presently, one of the most reliable ways of generating a large EMP burst is through a nuclear detonation at high altitude. At 20–40km above ground level, the EMP burst of a nuclear weapon would be very large; at 400km above ground level, this EMP could be as large as the continental United States (Federation of American Scientists, 1998). Ambitious countries may redouble their efforts to gain control of nuclear weapons as a form of defense against HANDs, which could be devastating to international security.

A different type of arms race may occur in terms of cyberwarfare. As HANDs would rely, presumably, on wireless connections, the invasion of wireless networks will become a tool for enemy combatants facing a force of HANDs. Such invasion may merely result in disconnecting an operator from her vehicle, but may also involve hijacking the machine, or sending malicious or lethal signals back to an operator. This will in turn motivate increasing advances in cryptography and wireless security that resists these attacks and advancements in the ability to track incoming signals and detect these "cybercombatants."

In this case, asymmetric war would experience a paradigm change, from small bands of insurgents conducting guerilla war, to a split force, of one set of "conventional combatants within a warzone," while another set of cybercombatants exist outside the conventional warzone attempting to invade the information networks being run by military organizations using HANDs. This could turn asymmetric warfare into global conflict, as the legitimate combatants of a war are stationed equally far from the warzone in question as the neurowarriors. This may, as I have noted, render certain targets permissibly attackable by neurowarriors, but the consequences of this are rather different from conventional war, or even conventional asymmetric war—neurowarriors fight a war with a particular objective at location A from location C while they are attacked from location D.

Finally, there may be a redoubled effort for weaker countries to build and implement AMRs. By using autonomous robots, unscrupulous armies may disregard problems associated with collateral damage or responsibility, and rather use robots to attack HANDs where they cannot reliably or safely resist such efforts. This could become incredibly problematic, for reasons Sparrow has discussed regarding the ethical problems with autonomous robots in war (Sparrow, 2008).

Alternately, the presence of a HAND may indeed bring about risk-free warfare, and following Khan, the slide of military action into policing (Khan, 2002, p. 7). This brings with it many problems, not least of which are the international legal ramifications of international policing efforts. I will not discuss presently what this might mean for international sovereignty. It remains to be seen whether, in our international climate, international policing of this type is plausible, morally justified, or feasible. Whoever controls such a technology would have a terribly advanced form of power, not only removing them from being able to contested against militarily, but with the added ability to enforce their agenda without the ability to be resisted properly. The enforcement of a criminal justice system has similar conditions to those of war, including necessity and proportionality requirements (Kleinig, 2010). However, the justification for use of lethal force rests on a different basis in the pursuit of criminal justice, as the purpose of force is to bring individuals or groups to trial, not to complete military objectives. This would, if it was adopted successfully, potentially be a very positive step for the use of force in an international context, but could also be highly problematic if abused or managed poorly.

Finally, we might think that we need to keep embodied troops on the ground. We may decide, through the use of arms control treaties and other mechanisms that HANDs are only to be used in nonlethal or nonsoldier roles: scouts, defensive roles ("robotic shields"), policing roles, medics, and so on. This would, subject to the appropriate international agreement and oversight, be a positive outcome. HANDs in these roles are potentially useful and beneficial, as well as for their civilian uses such as restoring mobility to motor-impaired individuals.

Alternately, this may create some form of perverse incentive for military forces. Say that international law forbids the use of HANDs with lethal intent unless other ground forces are threatened. This would conceivably promote the use of HANDs in a role where small numbers of ground forces are used effectively as bait: in the resulting defence of these combatants, the HANDs are permitted to defend their compatriots with their lethal munitions, though in a patently unfair conflict.[5]

If we are to use HANDs, then we must also engage the ethics of nonlethal weapons with an increased vigour. HANDs, if they are to be used widely, will most likely need to use a variety of nonlethal weapons to subdue individuals in pursuit of their objectives. At present there are many nonlethal weapons, but the statuses of these devices are not confirmed for their efficacy, safety, and what entails "nonlethal." Moreover, reliable ways of taking prisoners would have to be taken into account once hostile individuals were subdued.

Conclusion

HANDs solve a number of problems related to the use of military robots either autonomously or remotely. However, in doing so they create a situation in which there is no longer a strong rationale for a soldier to fire on the enemy. This may lead to a number of different outcomes: an increasing asymmetry in war forcing increased terrorism, a new arms race, or a change in international armed conflict from a military to a policing model. HANDs are in the future, to be sure—they are the next step in unmanned military technologies. At this time I cannot make comment on which outcome will occur, but more debate needs to occur so that the highly destructive outcomes of this trend towards risk-free warfare are averted.

Bibliography

Arkin, R.C. (2007), *Governing Lethal Behavior: Embedding Ethics in a Hybrid Deliberative/Reactive Robot Architecture* (Technical Report GIT-GVU-07-11 for US Army. Mobile Robot Laboratory, College of Computing, Georgia Institute of Technology, Available online at http://www.cc.gatech.edu/ai/robot-lab/online-publications/formalizationv35.pdf, accessed August 18, 2010).

Downes, C. (2004), "'Targeted Killings' in an Age of Terror: The Legality of the Yemen Strike," *Journal of Conflict and Security Law*, 9:2, 277.

Etoh, M. and Yoshimura, T. (2005), "Advances in Wireless Video Delivery," *Proceedings of the IEEE*, 93:1, 111–22, doi: 10.1109/JPROC.2004.839605.

Federation of American Scientists (1998), *Nuclear Weapon EMP Effects* (Available at http://www.fas.org/nuke/intro/nuke/emp.htm, accessed August 18, 2010).

Garreau, J. (2006), *Radical Evolution: The Promise and Peril of Enhancing Our Minds, Our Bodies—and What It Means to Be Human* (New York: Broadway Books).

---

5 Many thanks to Adam Henschke for bringing up this problem in conversation.

Hurka, T. (2008), "Proportionality and Necessity," in May, L. (ed.), *War: Essays in Political Philosophy* (New York: Cambridge University Press).

Khan, P.W. (2002), "The Paradox of Riskless Warfare," *Philosophy and Public Policy Quarterly*, 22:3, 2–8.

Killmister, S. (2008), "Remote Weaponry: The Ethical Implications," *Journal of Applied Philosophy*.

Kleinig, J. (2010), *Ethics and Criminal Justice: An Introduction* (New York: Cambridge University Press).

Lauder, S. (2010), "Trial of Bionic Eyes within Three Years," ABC News (March 30. Available at http://www.abc.net.au/news/stories/2010/03/30/2860256.htm, accessed August 20, 2010).

Licklider, J. (1960), "Man-Computer Symbiosis," HFE-1 IEEE *Transactions on Human Factors in Electronics*, 4, 4–11.

McMahan, J. (2009), *Killing In War* (Oxford: Oxford:University Press).

Rodin, D. (2002), *War and Self-Defense* (Oxford: Oxford University Press).

Scerri, P., Von Gonten, T., Fudge, G., Owens, S., and Sycara, K. (2008), "Transitioning multiagent technology to UAV applications," in *Proceedings of the 7th International Joint Conference on Autonomous Agents and Multiagent Systems*: Industrial Track (Estoril, Portugal), 89–96.

Scheutz, M, Schermerhorn, P. and Bauer, P. (2005), "The utility of heterogeneous swarms of simple UAVs with limited sensory capacity in detection and tracking tasks," *Swarm Intelligence Symposium*, SIS 2005. Proceedings 2005 IEEE (June 2005), 257–64.

Santhanam, G., Ryu, S.I., Yu, B.M., Afshar, A. and Shenoy, K.V. (2006), "A high-performance brain-computer interface," *Nature*, 422, 195–8.

Sparrow, R. (2009), "Predators or plowshares? Arms control of robotic weapons," *Technology and Society Magazine*, IEEE, 28:1 (Spring), 25–9.

Sparrow, R. (2007), "Killer Robots," *Journal of Applied Philosophy*, 24:1, 62–77.

Tybaji, A. (2007), "Unique problems associated with UAV employment," *Flying Safety* (May 2007).

Walzer, M. (2000), *Just and Unjust Wars: A Moral Argument With Historical Illustrations* (New York: Basic Books).

Chapter 7

Robotic Weapons and the Future of War

Robert Sparrow

Introduction

Robotic weapons are likely to play a central role in future wars (Featherstone 2007; Graham 2006; Hanley 2007; Hockmuth 2007; Peterson 2005; Scarborough 2005; Singer 2009).[1] The perceived success of the Predator uninhabited combat aerial vehicle (UCAV) in Iraq and Afghanistan has resulted in military forces around the world rushing to develop and field unmanned systems (UMS) (Hockmuth 2007; Office of the Secretary of Defense 2005). Existing UMS are, for the most part, operated by a remote operator but some also make use of sophisticated artificial intelligence technology in order to provide the weapon with some capacity for autonomous action (Excell 2007; Braybrook 2007). In the future, the scope of autonomous action allowed to UMS may increase, with the possibility that Autonomous Weapon Systems (AWS), capable of being deployed to achieve particular military objectives without human supervision, will eventually be developed (Office of the Secretary of Defense 2005, p. 52; Singer 2009, pp. 123–34).

It is clear that the use of UMS in warfare will generate a number of challenging ethical issues—more than can be dealt with in a single chapter.[2] I have written about the difficulties involved in attributing responsibility for deaths resulting from the actions of autonomous weapon systems (Sparrow 2007a), and the possibility that robotic weapons will lower the threshold of conflict, elsewhere (Sparrow 2009a). I have also addressed the issues that may confront engineers and other researchers involved in the design of these systems (Sparrow 2009b). In this chapter I want to examine a number of issues raised by the development and deployment of robotic weapons that should be of special interest to those writing and thinking about the ethics of war more generally. That is, I have here chosen to discuss those issues

1   The research for this chapter was supported under the Australian Research Council's Discovery Projects funding scheme (project DP0770545). The views expressed herein are those of the authors and are not necessarily those of the Australian Research Council. I would like to thank Neil McKinnon for assistance with research for this chapter. I would also like to thank Jessica Wolfendale, Tony Coady, Linda Barclay, Jim Sparrow, and Ron Arkin for comments and discussion that have improved this chapter.

2   Other discussions include: Asaro 2008; Krishnan 2009; Borenstein 2008; Sharkey 2009; Singer 2009.

that raise the most pressing philosophical problems and/or where the impacts of robotic weapons on the nature of armed conflict are likely most profound.

With any discussion of the ethical issues raised by a given technology, there inevitably arises the question of whether the issues are in fact unique to the technology under discussion or whether they are associated with the technology's predecessors. Is the "new" technology really that novel or is it just an extension of existing and historical technologies? (Sparrow 2007b). This question arises with especial force regarding military robotics because there is a long history of the use of machines in warfare. Both of the features of robots that are responsible for the majority of qualms about their military applications—that they allow "killing at a distance" and that the operations of the machine may determine who lives or dies—are shared with other technological systems. It might therefore be argued that military robotics is undeserving of being singled out for focused ethical examination.

For current purposes, I wish to remain agnostic on the question of the extent to which the ethical issues raised by robotics are unique to robots. While many of the issues I discuss below are familiar, others are less so. In particular, some of the issues raised by the (hypothetical) development of *autonomous* weapon systems seem to have relatively few precedents in existing weapons. More generally, recent and likely future developments in robotics exacerbate and bring into stark relief issues that arise elsewhere in less dramatic forms. In any case, ethical issues do not need to be novel to be important. Given that robotic weapons are coming into widespread use for the first time now, it is appropriate—indeed, necessary—to give whatever ethical issues they raise serious consideration. It is to this task I now turn.

The Ethics of Killing-at-a-distance

Perhaps the first issue to arise in any discussion of robotic weapons is the ethics of killing at a distance. The pilots who operate Predator and Reaper are safely ensconced thousands of kilometres away from the people they kill. The drones they operate fly so high and so quietly that the people that they target often have no idea that they are there until their missiles arrive. It is hard *not* to wonder about the ethics of killing in this fashion. In particular, is hard not to think that by distancing warfighters from the consequences of their actions, both geographically and emotionally, UMS will make it more likely that the operators of robotic weapons will make the decision to kill too lightly (Bender 2005; Ulin 2005).

Ultimately, whether the use of UMS will contribute to more or less ethical behaviour by their operators is an empirical matter that can only be resolved by careful study over an extended period of time. However, there are number of reasons to think that robotic weapons are less problematic in this regard than first appears.

To begin with, any analysis of the impact of robotic weapons on the behaviour of those operating them must keep in mind the character of the weapons these systems replace. Warfare has involved "killing at a distance" ever since the invention of the sling and the spear. Modern warfare includes many weapon systems that allow their operators to kill people they have never met and that distance those that operate them from the destruction that they wreak (Sullivan 2006). The operators of UAVs, at least, are *better* situated to learn about the consequences of their actions than, for instance, artillery gunners, bombardiers, or other warfighters who kill at a distance using existing weapons systems. The pilots who fly UAVs are able—indeed, are usually required—to conduct a battlefield damage assessment which exposes them to (some) knowledge of the consequences of their actions.[3] One might therefore hope that these pilots will take the decision to kill at least as seriously as their comrades. Other remotely operated systems (ground vehicles, submersibles, surface vehicles, etc) may not offer as good a perspective on the battlespace as UAVs but will still usually do better than comparable systems without operators. It is also worth noting that the telemetry from remotely operated weapon systems will usually be recorded and be available for review, which may also serve as a significant incentive to obey the laws of war.

It must also be acknowledged that close proximity between enemies is far from being a panacea against illegitimate killing. Some of the worst atrocities of modern times have been carried out at close quarters by men armed with rifles and machetes. Any attempts to evaluate the impact of UMS on the extent to which those fighting wars are willing to kill will need to be made in the context of a realistic appraisal of ways in which people behave in more familiar and less high-tech forms of conflict (Arkin 2009, 30–6). If the appropriate comparison is Kosovo or Rwanda then the argument that UMS will make killing more likely may be hard to establish.[4]

---

3   However, whether the sensor systems of UAVs are capable of communicating the reality of these consequences in such a way that knowledge of this reality might inform the future moral character of those who kill using these systems, is less clear. While video taken from 15,000 feet may be adequate to determine whether tanks or artillery have been destroyed and people killed, it may not be enough to convey the *moral* reality of killing (Cummings 2004). The sorts of experiences that are essential to understanding what it means to kill may be too "embodied" and subtle to be transmitted via video (Mander 1978). For further discussion, see Sparrow (2009b).

4   My thanks to Linda Barclay for encouraging me to admit the force of this objection.

How will Robots Transform the Requirements of Discrimination and Proportionality?

An issue which has received some useful discussion in the literature already is the way in which the development of UMS may affect the proper interpretation of the principle of discrimination within the law of war (Meilinger 2001). This principle, which is a central part of the just war doctrine of jus in bello, requires that warfighters not directly target non-combatants and also take reasonable steps to avoid non-combatant casualties (Lee 2004). What is reasonable in this context depends in part on the technical means available to the warfighter (Dunlap 1999, Schmitt 2005). UMS can greatly increase the precision with which weapons can be targeted by making available real-time targeting information (Kenyon 2006, Meilinger 2001; Tuttle 2003). They may also be capable of delivering weapons in such a fashion as to ensure maximum accuracy in circumstances where it would be too risky for a manned system to attempt to do so. The more it becomes possible to identify and destroy specific targets using UMS, the more commanding officers may be obligated to make use of these systems. An unintended consequence of the development of these more versatile systems may be that commanders eventually face new restrictions in their use of older generations of weaponry.

The requirements of proportionality may also be affected by the development of UMS. Under jus in bello warfighters are only to use force proportionate to legitimate military goals. Again, what is proportionate use of force will depend, in part, on the range of possible applications of force that are available to achieve the required goal. In so far as UMS allow warfighters to apply destructive force more precisely they will also allow them to use a smaller amount of force. The destruction involved in the use of other weapon systems may eventually come to appear disproportionate by comparison.

The development of UMS may affect the requirements of proportionality via another route. The degree of force that it is permissible to use depends on the nature and importance of the military goal that is being pursued. When warfighters' lives are at stake, it is acceptable to use a large amount of destructive power to preserve them. However, in operations involving (only) UMS, enemy combatants will pose no threat to the lives of their operators. Even if this does not entirely undercut the justification for the use of lethal force, it may nevertheless affect our judgements about the level of force that is proportionate in response to an attack on a UMS. Obviously, the degree to which the absence of a threat to the operator diminishes the amount of violence that constitutes a proportionate response will depend on the particular circumstances in which an operator of the UMS might be considering using lethal force and the nature of the conflict in which the system is engaged. While enemy combatants may not pose an urgent threat to the operators of the UMS, they may pose such a threat to other warfighters. Similarly, while the destruction of a UMS might not harm its operators, it may threaten lives elsewhere in the battlespace or at another time when the UMS would otherwise have played a role in reducing the threat to combatants. If we think of enemy warfighters in general as posing a threat—as I will discuss further below—then the operators of UMS may be justified in using whatever amount of force is necessary to kill them where-ever they encounter them. However, if our sense of a proportionate response is linked to the threat posed by a particular unit in particular circumstances, then the proportionate response from a UMS to a military threat may be significantly different to that of a manned system.

Importantly, the development of UMS also opens up the possibility that *new* targets may become available for legitimate attack (Hambling 2007). In so far as they make possible more discriminating and proportionate uses of lethal force, UMS may allow warfighters to attack targets that previously could not be attacked without risking excessive casualties (Arkin 2009; Dunlap 1999; Meilinger 2001; Metz 2000; Office of the Secretary of Defense 2005; Schmitt 2005). Of course, this also means that military forces may attack *more* targets, which in turn may have implications for the likelihood that military campaigns will produce civilian casualties.

The implications of the development of UMS for the requirements of the principles of discrimination and proportionality are thus more complex than first appears. It is clear, however, that the proper interpretation of the principles of discrimination and proportionality will be affected by the evolution of robotic weapons. Changes to the interpretation of these principles may have significant implications for the future of war.

Autonomous Weapon Systems and Jus in Bello

There is another important question that arises about the principles of discrimination and proportionality if *autonomous* weapon systems come into widespread use. That is, if decisions about targeting and the use of lethal force are eventually handed over to machines.

Initial discussions of robotic weapons by those advocating their adoption were careful to insist that the decision to take human life would never be made by a machine. While a robot might identify and track a target, the decision to attack it would always be made by a human being. Thus, there would always be a human being "in the loop." More recent discussions have come to endorse a significantly looser idea of human control over the operations of the weapon system as a whole rather than each targeting decision it makes—a human "on the loop" rather than "in the loop" (Arkin 2009; Singer 2009, pp. 123–34). With the benefit of hindsight, it is reasonably clear that the previous insistence that machines will never be allowed to make a decision about targeting human beings was, at the very least, naive and probably disingenuous (Singer 2009, pp. 123–34). The logic of the dynamic that has produced these weapons pushes towards the development of autonomous weapon systems (Adams 2001). Now that it has become possible to take the pilot out of the weapon platform, the question naturally arises whether it might be possible to take them out of the system altogether. The tempo of battle in air combat is already so high that it is difficult for human pilots to make decisions in the time available to them. Consequently, humans already rely upon automated systems such as threat assessment displays to make life or death decisions. Eventually the tempo of battle may become so high that it is impossible for human beings to compete with an automated system that has the authority to fire weapons; at this point, the temptation to hand over control of the entire platform to an automated system will be almost irresistible. Moreover, the communication infrastructure necessary to maintain the link between the operator and the weapon is an obvious weak point in the operations of the existing drones. Competition for satellite bandwidth already places significant limits on the number of drones that can operate over Iraq and Afghanistan. An enemy with the capacity to do so would almost certainly try to destroy or disrupt this communication infrastructure in the opening stages of a conflict. Again, the obvious response is to eliminate the need for a human "in the loop" altogether. Given these pressures, it is important that we think seriously about whether it is possible for robots to obey the dictates of jus in bello.

Ron Arkin, Professor of Robotics at Georgia Tech, is already advocating the development of weapon systems capable of operating in a fully autonomous mode (Arkin 2009). Designing robots that are capable of discriminating between legitimate and illegitimate targets under the laws of war involves formidable technical challenges. It is difficult enough for a robot to distinguish a person from their surroundings, let alone a person in uniform from a person not in uniform or—most importantly—a combatant from a noncombatant. Nevertheless, Arkin argues that engineers should take up this challenge, even to the point of aiming to develop an "ethical governor" that can sit within the decision-making processes of military robots in order to allow these machines to distinguish between ethical and unethical killing (Arkin 2009, pp.127–53).

I am inclined to think that this project underestimates the extent to which the proper interpretation of the principles of jus in bello is context dependent. Determining the application of concepts that are central to the principles of jus in bello, such as "proportionality," "threat," and "reasonableness," requires judgement at an extremely high level of abstraction. Whether or not an armed tribesman in Afghanistan, for instance, is a legitimate target for attack by US forces in the country may depend upon his location, his past movements, his tribal and political affiliations (if known), the history of the local area, and the disposition of other forces and civilians nearby. For a machine to gather the relevant information, let alone represent or process it, would require a computer of unprecedented power and sophistication. Even in more conventional warfare, not all enemy soldiers in uniform are combatants—as they may have indicated their desire to surrender or be so incapacitated by wounds as to no longer represent a threat. Indeed, morally speaking, even uniformed soldiers holding weapons may not be legitimate targets in some circumstances, for instance, if their deaths would be disproportionate to the military goal that would be achieved by killing them. The extent to which judgements about discrimination and proportionality are sensitive to context makes it hard to imagine how any system short of a genuine artificial intelligence might even begin to apply the principles of jus in bello.[5] For this reason, I remain cynical about the likelihood that automated systems will ever be good enough to be entrusted with decisions about who to attack.

Arkin has a powerful reply available to him at this juncture, which is to point out that it is not necessary to achieve 100 percent accuracy in discriminating between legitimate and illegitimate targets but only to achieve a level of discrimination comparable to that of human warfighters. Moreover, as Arkin (2009, pp. 30–6) observes, human beings are not that good at applying the principles of jus in bello. In practice, human beings often fail to discriminate between legitimate and illegitimate targets in wartime, for at least two reasons. First, discrimination is difficult. Decisions often need to be made quickly, under considerable pressure, in chaotic circumstances, and often on the basis of limited information and resources. Little wonder, then, that even the most conscientious of warfighters sometimes make mistakes. Second, even where they are capable of distinguishing between these categories, they may be insufficiently committed to making the distinction. Obeying the dictates of the principles of jus in bello will sometimes require significant personal sacrifice from warfighters in terms of the risks to which they must subject themselves. If they hesitate too long before firing on a target or fail to fire on a target about which they are unsure, this may lead to their being killed. Similarly, if they refrain from firing on enemy combatants for fear of causing disproportionate civilian casualties, this may allow the enemy to escape or to continue firing on them. There are also significant social and psychological pressures on warfighters, which may encourage them to come to view all of the members of an enemy political, ethnic, or racial group as a threat to them or otherwise undeserving of concern and respect. All too often, these factors combine to produce an attitude that might be expressed colloquially as "kill 'em all, let God sort them out."

Given that human beings tend only to apply the rules of jus in bello imperfectly in practice, it is not inconceivable that in some domains automated systems will prove more reliable than human warfighters at distinguishing legitimate from illegitimate targets. This is even likely in applications, such as air combat or missile defence, wherein the temple of battle is such that human beings *cannot* make decisions about firing (or not) in time to be effective (Adams 2001).

However, this way of understanding the principle of discrimination treats it as though it were merely a technical challenge—a problem of pattern

---

5   In fairness to Arkin, he is aware of this problem and suggests that autonomous weapons should be designed to call for assistance from a human supervisor when they encounter a situation that requires judgement of this sort (Arkin 2009, 177–209). This presumes, of course, that reliably determining when one is confronting an ethical dilemma is not also a task requiring a high level of context sensitive abstract reasoning. Moreover, any need for access to human supervision substantially undermines the purported advantages of autonomous weapon systems and may be expected to generate real tension with the imperatives that are driving the development of such systems.

recognition—of the sort that roboticists routinely confront. There may be more to the principle of discrimination than this.

In a famous and influential article on the moral foundations of the principles of jus in bello, Thomas Nagel (1972) argued that the force of these principles could only be explained by the idea that they are founded in absolutist moral reasoning. Consequentialist moral philosophy is shaky grounds upon which to found the injunctions of just war theory, as it is all too easy to imagine circumstances in which a clearheaded calculation of consequences speaks against respecting the rules of war. Thus Nagel offers an account of the key injunctions of jus in bello by way of a (essentially Kantian) principle of respect for the moral humanity of those involved in war. He argues that even during wartime it is essential that we acknowledge the personhood of those with whom we interact and that,

> whatever one does to another person intentionally must be aimed at him as a subject, with the intention that he receive it as a subject. It should manifest an attitude to him rather than just to the situation, and he should be able to recognise it and identify himself as its object. (p. 136)

Another way of putting this is that we must maintain an "interpersonal" relationship with other human beings even during wartime. Obviously, if this principle is to serve as a guide to the ethics of war—rather than a prohibition against it—the decision to take another person's life must be compatible with such a relationship (Nagel 1972, p. 138).

Nagel's discussion suggests another interpretation of the principle of discrimination, in terms of what is required to extend people the sort of respect that is appropriate to making the decision to end someone's life. Discrimination is the act of establishing the interpersonal relationship with a possible target of attack and responding to relevant features of this person, which may or may not justify attacking them with lethal force. If this account of the nature of discrimination is correct then merely showing that a weapon system reliably attacks only legitimate targets will *not* be enough to argue that it is discriminating in the appropriate sense.

Indeed—at first sight at least—it appears that autonomous weapons would *not* be capable of discrimination on this account because they are not capable of establishing the appropriate interpersonal relationship with those they target. When a robot does the killing there is no one who cares whether those killed are combatants or non-combatants or what it is that combatants are doing that makes them legitimate targets. According to this way of thinking, placing the lives of our enemies at the mercy of a machine seems to neglect and devalue their humanity in quite a profound way; we do not even grant them the courtesy of deciding whether they are worth killing.

However, this presumes that it is the robot that should be manifesting the respect. Unless the machine is morally autonomous—which would require it, at the very least (Sparrow 2004), to be a genuine artificial intelligence—it might be argued that the proper place to look for the required attitude is rather in the person who ordered the deployment of the AWS. Many existing weapons systems— including those that AWS might plausibly replace—seem to involve only a very abstract and mediated relationship between those who use them and those they are trying to kill. If there can be an "interpersonal" relationship between a bombardier and the enemy combatants they target thousands of feet below, it is not clear why a similar relationship could not exist between the person who orders the deployment of an autonomous weapon system and the people who are killed by that system. We might, then, think of an autonomous weapon as being merely a better (smarter?) "smart bomb." Given the choice, enemy civilians might well prefer that they face robotic weapons rather than bombing from a B52 or even than gunfire from nervous 19-year-old Marines. Insofar as the principles of jus in bello sometimes also protect the lives of enemy combatants it is even possible that enemy combatants would prefer to be fighting robotic weapons that reliably implement these principles rather than human soldiers or other technologies that do not. Thus, it seems arguable that the use of autonomous weapons *is* compatible with a respect for the humanity of those being killed.

How we think about the morality of autonomous weapons will depend in part, then, on how plausible it is to understand them as simply a means whereby the person who commands their use kills people. If we think of AWS as transmitting their commanding officer's intentions, then there will be no question as to whether the robot itself is discriminating in the right way to satisfy the principle of discrimination. However, to the extent that we think of the machine itself as choosing its targets, then even a AWS that is better than a human being at selecting the "right" targets may not be capable of discrimination according to the analysis I have developed here. Which of these interpretations is more plausible is, I think, likely to become one of the key questions about the ethics of the deployment of autonomous weapon systems.

There is one further factor to consider. As Nagel observed in his original paper, how we express respect is, in part, a matter of covention. Thus, for instance, what counts as desecrating a corpse will be determined by conventional social meanings and practices, etc. Thus, prevailing social understandings about what is required to respect one's enemy have some status simply by virtue of being widely shared. Moreover, the military's reluctance to admit that machines will eventually be given the capacity to choose their own targets and decide when to release their weapons reflects, I think, their awareness of the strength of the public's feeling about the idea that machine should have the right to decide who lives or dies. Giving an autonomous weapon systems the power to kill seems a bit too much like setting a mousetrap for human beings; to do so would be to treat our enemies like vermin.

The requirements of convention, then, may well swing the balance of considerations towards the conclusion that even (hypothetical) autonomous weapon systems that are capable of reliably distinguishing between legitimate and illegitimate targets under the laws of war would fail to extend the appropriate respect to those they might kill. Of course, such a "conventional" account is vulnerable to the possibility that social understandings may change, especially as the public gets more used to the idea of robotic and autonomous weapons. Moreover, it must be acknowledged that conventions are shallow foundations for moral principles that must stand up to the temptation to bend or break them that almost inevitably occurs in war. Thus while I think the balance of considerations favors the conclusion that the use of autonomous weapon systems would be unethical by virtue of its failure to respect the personhood of their targets, this conclusion must remain a tentative and provisional one, pending further philosophical discussion and investigation.

Will Robots Undermine "Warrior Virtues"?

In many—if not most—societies, military service—and sometimes even war itself—is exalted because of its putative role in shaping character and developing particular virtues. War and military training provide an opportunity for the exercise and development of physical courage, fortitude, loyalty, willingness to sacrifice for the sake of the community, and even love and compassion. The idea that there is a connection between war and virtue also plays an important role in the culture of military organisations, which typically promote a conception of a professional warfighter who is supposed to cultivate and demonstrate "warrior virtues" (Olsthoorn 2005; Robinson 2007; Toner 2000; United States Department of the Navy 2007; Watson 1999). These virtues both serve the pursuit of victory in war and place limits on the ways in which wars are fought. "Good warriors" do not abandon their comrades in battle, nor do they massacre civilians (Silver 2006).

Taking humans "out of harm's way" by allowing them to fight remotely will also remove them from the opportunity to exercise those virtues that are currently constitutive of being a good warrior. Most dramatically, it does not seem possible to demonstrate courage while operating a UMS from outside of the theatre of battle (Fitzsimonds and Mahnken 2007, pp. 100–101; Shachtman 2005). It is also hard to see how their operators could demonstrate a willingness to sacrifice themselves for the good of the unit. It will still be possible for operators to work hard and above and beyond the call of duty but it will not be possible for them to demonstrate any of those virtues that require the existence of serious negative consequences or physical threats as conditions of their possibility. More controversially, the limitations of UMS may make it extremely difficult for the operators to demonstrate love and compassion through their operations. Again, while they might go "above and beyond" to save the life of a particular comrade (or non-combatant) to such an extent that we might normally be inclined to attribute their actions to love or compassion, the geographical and psychological distance between the operator and the UMS means that these actions may nonetheless fail to demonstrate these virtues.

The loss of the opportunity for warfighters to demonstrate virtues may, of course, prove an acceptable price to pay to preserve their lives and limbs. Similarly, warfighters may be happy to live without a knowledge of what it means to kill. What is less clear is whether military culture and military organisations can continue to function, in their current form at least, if it no longer becomes plausible to posit a connection between war and the character of those who fight it.[6] The development of UMS may thus pose a serious threat to the self-conception of the military and perhaps even to the long term survival of the military services in their current forms.

Will Robot Weapons Make War Immoral?

The development of remotely operated weapon systems that would allow their operators to kill others without risk to themselves poses a serious challenge to the justification of war itself. Yale University Professor Paul Kahn has argued that unless it involves risk to both parties *war* is unethical (Kahn 2002). Where the asymmetry between opposing forces is large enough such that one party faces little or no risk, the moral exercise of force by the state possessing superior military might must take on the character of *policing*.

According to Kahn, the primary difference between these distinct modes of engagement concerns the categories of people who may legitimately be targeted with lethal force and the justification for doing so. In war it is legitimate to target and kill enemy combatants—those who are actively involved in the business of warfare (Walzer 2000, pp. 144–6). It is permissible to kill (some) non-combatants in the course of attacks on combatants as long as the non-combatants were not themselves directly targeted and reasonable care was taken to avoid and/or minimise non-combatant fatalities. In policing, on the other hand, it is permissible to kill only those who are responsible for serious violations of human rights and even then only in the course of preventing serious rights violations and/or trying to bring the perpetrators of such violations to justice. It is almost never morally permissible to kill persons who are not responsible for such rights violations in the course of policing operations (Kahn 2002).

The justification of the use of force in policing therefore proceeds by reference to the concepts of individual guilt and responsibility. The justification for killing in war is more complex and controversial. A full investigation of this topic is beyond the scope of this chapter. My ambition here is more limited. Following Kahn, I want to bring out the ways in which our intuitions about legitimate killing in war rely on the presupposition that opposing combatants are mutually at risk.

---

6   Fitzsimonds and Mahnken (2007) note that their enquiry, in the course of their survey of the attitudes of officers to UAVs, about whether the operators of UAVs should be eligible to receive the Distinguished Flying Cross for heroism in aerial flight "elicited a number of written comments ..." and that "One respondent termed the question itself 'incredibly disturbing'" (p. 101), suggesting that contemporary operations of UAVs are already unsettling the self-conception of some military personnel.

To begin with, note that there are no morally relevant features that distinguish those it is permissible to directly kill (combatants) and those it is not (non-combatants) other than their membership in these groups. Combatants do not "deserve" to die any more than noncombatants. Many combatants do not support the war effort. Many non-combatants are fierce supporters of the war. Where conscription is practised, individuals may not even have chosen their status as combatants (Kahn 2002). Instead, the distinction between combatants and non-combatants is itself a product of war, which is a relationship between states (or state-like actors) (Sparrow 2005). It is war that makes combatants enemies and confers on them a licence to kill each other. Members of the armed forces of one state do not possess a licence to kill members of the armed forces of another state with which the state they serve is not at war.

Importantly, the fact that two states are at war does not confer on *all* their citizens the right to kill the "enemy." It is only those in the service of the state in certain roles that lose their right not to be killed and who gain a licence to kill those of their enemies who serve in similar roles. Indeed, there is a sense in which non-combatants are not each others' "enemies" at all (Lee 2004, p. 238). For instance, if an Australian and a Vietnamese civilian met during the American War in Vietnam and one killed the other, it would be an inadequate justification to explain that they were enemies because they were "at war." What makes combatants enemies in a strong sense is that through their activities in the pursuit of the state's military goals they are also working to kill each other (Murphy 1985). They stand in a relationship of "mutual threat" and their "licence to kill" stems from their right to self defence (Fullinwider 1985; Kahn 2002). Our ethical practices concerning the treatment of combatants who have been wounded or who have surrendered, in particular, strongly suggest that threat plays a crucial role in underpinning the right of combatants to kill other combatants. Surrender restores to combatants their right not to be killed. Similarly, because they have ceased to be a threat, severely wounded warfighters who have been overrun in battle are not legitimate targets in war.

However, if threat plays a central role in establishing warfighters "licence to kill" then, as Kahn argues, should the armed forces of one side of a conflict be rendered incapable of posing a threat to the other, then the warfighters of the more powerful side will no longer have a justification for targeting the armed forces of the other. Instead, the moral permissions and obligations of war are replaced by a more limited right to self defence alongside the permissions and demands of policing.

It is important to acknowledge that the relevant asymmetry is that which may exist between military forces considered as a whole. Kahn's paper neglects this aspect of the argument and, by emphasising that combatants' permission to kill each other is founded in their individual right to self defence, risks conveying the impression that it is the immediate threat to their person that is relevant. However, warfighters routinely kill others who pose no immediate threat to their persons in war, as, for instance, when they shell enemy positions over the horizon or bomb munitions factories from a great height. Any plausible argument in support of their permission to do this cannot be founded on their right to defend themselves against *specific* threats. Instead, the threat that warfighters on opposing sides pose to each other and which justifies their licence to kill is a general one. *All* of the combatants on one side are a threat to *all* of those on the other. This feature of war is highlighted, in the course of satirising it, in the following section of dialogue from Joseph Heller's *Catch 22*.

> 'They're trying to kill me,' Yossarin told him calmly.
> 'No one's trying to kill you,' Clevinger cried.
> 'Then why are they shooting at me?' Yossarian asked.
> 'They're shooting at *everyone*,' Clevinger answered. 'They're trying to kill everyone.'
> 'Who's they?' he [Clevinger] wanted to know. 'Who, specifically, do you think is trying to murder you?'
> 'Every one of them,' Yossarian told him. (Heller 1994, p. 19)

However, that "they" are "trying" to kill me is not enough to justify my killing them in self defence; they must also have a realistic chance of doing so, such that they constitute a threat to my safety. If the military asymmetry between two sides to a conflict becomes large enough then the forces of the weaker side may not pose any such threat.

The presence of robotic weapons in a nation's armed forces is neither necessary nor sufficient to establish asymmetry of this sort. Indeed, Kahn makes his argument without specific reference to UMS. What matters is the extent and nature of the asymmetry, not how it is achieved. However, by opening up the possibility that all human beings could be removed from the direct line of sight of the enemy and that indirect fire threats could be destroyed as soon as they appear in the battlespace, robotic weapons make it much more likely that the required asymmetry might be established. The only other foreseeable circumstances in which the armed forces of one side might pose no significant threat to the other might be a military campaign conducted entirely from the air.[7] In either circumstance, if the argument that I am exploring here is to have any application it must be the case that there is some threshold of risk that establishes that the asymmetry between contesting forces justifies only policing and not war rather than the *complete* absence of risk. It is hard to imagine any conflict in which enemy combatants pose *no risk at all* to friendly forces. However, it *is* possible to imagine a circumstance in which the forces of one side presents so little threat to those of the other that warfighters on the superior side are not justified in directing lethal force against the warfighters of a conscript army merely because they are bearing arms.

---

7 This is the context in which Kahn (2002) developed his argument, discussing the NATO campaign over Bosnia. See also Kahn (1999).

It is possible, then, that future advances in the development of UMS may have the perverse consequence of undermining the justification of their use in war (or at least, war against certain sorts of enemies). Instead, robotic weapons would need to be used (much) more selectively, to prevent injustice and punish crime in the context of international policing operations.

Robotic Wars and the Future of Ethics

I have not attempted here to reach any overall conclusion "for" or "against" the development or application of military robots. This would be a futile task given the wide range of capacities of these systems and roles in which they might be used. Rather, my aim has been to draw attention to various implications of the development of robotic weapons for the future of war that I believe are significant for the ethics of war.

An important observation that emerges from this discussion is that it is far from obvious that all of the implications of the use of military robots make war "worse." The argument that they will make killing more likely by allowing "killing at a distance" is less powerful than one might have thought. Moreover, insofar as these weapons allow warfighters an increased ability to discriminate between legitimate and illegitimate targets, they may both reduce the number of noncombatant casualties and also increase the demandingness of the principles of jus in bello. However, these possibilities need to be weighed against the possibility that the availability of UMS will make the use of force more likely, both at a tactical and strategic level (Sparrow 2009a). Ultimately, the impact of robotic weapons on how wars are fought is, of course, an empirical matter. Yet, given that controversy has erupted—and may be expected to continue to embroil—the introduction of these weapons, it is worth observing that the matter is more open than first appears.

I have also highlighted a number of other issues that are worthy of further investigation, both empirical and philosophical. In particular, there are reasons to expect the introduction of robotic weapons to impact on military culture in ways that might be extremely significant for the extent to which they encourage—or even allow—ethical behaviour. This is therefore an important topic for further study by those with skills to investigate it. The introduction of robotic weapons also raises difficult philosophical questions about the proper way of understanding the principles of jus in bello as applied to robotic—and especially autonomous—weapon systems and the ethics of "riskless warfare." I hope that my drawing attention to them in this context will allow other policymakers, ethicists, and philosophers to progress these important debates in order to keep pace with the ethical challenges that are likely to arise as we move into the era of "robotic wars."

References

Adams, T.K. (2001), "Future Warfare and the Decline of Human Decision-Making," *Parameters: US Army War College Quarterly* (Winter, 2001–2): 57–71.

Arkin, R.C. (2009), *Governing Lethal Behaviour in Autonomous Robots* (Boca Raton: CRE Press).

Asaro, P. (2008), "How Just Could a Robot War Be?" in P. Brey, A. Briggle and K. Waelbers (eds), *Current Issues in Computing and Philosophy* (The Netherlands: IOS Press), 50–64.

Bender, B. (2005), "Attacking Iraq, from a Nev. Computer," *Boston Globe*, April 3, A6.

Borenstein, J. (2008), "The Ethics of Autonomous Military Robots," *Studies in Ethics, Law, and Technology*, 2:1: Article 2. DOI: 10.2202/1941-6008.1036. Available at: http://www.bepress.com/selt/vol2/iss1/art2

Braybrook, R. (2007), "Drones: Complete Guide," *Armada International*, 31:3: 1–36.

Cummings, M.L. (2004), "Creating Moral Buffers in Weapon Control Interface Design," *IEEE Technology and Society Magazine* (Fall): 2–33, 41.

Dunlap, Jr., C.J. (1999), "Technology: Recomplicating Moral Life for the Nation's Defenders," *Parameters: US Army War College Quarterly* (Autumn): 24–53.

Excell, J. (2007), "Unmanned Aircraft: Out of the Shadows," *The Engineer*, January 18.

Featherstone, S. (2007), "The Coming Robot Army," *Harper's Magazine*: 43–52.

Fitzsimonds, J.R., and Mahnken, T.G. (2007), "Military Officer Attitudes Toward UAV Adoption: Exploring Institutional Impediments to Innovation," *Joint Force Quarterly*, 46: 96–103.

Fullinwider, R.K. (1985), "War and Innocence," in *International Ethics*, edited by C.R. Beitz, M. Cohen, T. Scanlon and A.J. Simmons (Princeton, N.J.:Princeton University Press).

Graham, S. (2006), "America's Robot Army," *New Statesman*, 135:4796: 12–15.

Hambling, D. (2007), "Military Builds Robotic Insects," *Wired Magazine*, 23 January. Available at http://www.wired.com/science/discoveries/news/2007/01/72543

Hanley, C.J. (2007), "Robot-Aircraft Attack Squadron Bound for Iraq," *Aviation. com*, 16 July.

Heller, J. (1994), *Catch 22* (London: Vintage).

Hockmuth, C.M (2007), "UAVs—The Next Generation," *Air Force Magazine*, February, 70–4.

Kahn, P.W. (2002), "The Paradox of Riskless Warfare," *Philosophy & Public Policy Quarterly*, 22:3: 2–8.

Kahn, P.W. (1999), "War and Sacrifice in Kosovo," *Philosophy & Public Policy*, 19 :2/3.

Kenyon, H.S. (2006), "Israel Deploys Robot Guardians," *Signal*, 60:7: 41–4.

Krishnan, A. (2009), *Killer Robots: Legality and Ethicality of Autonomous Weapons* (Farnham: Ashgate).

Lee, S. (2004), "Double Effect, Double Intention, and Asymmetric Warfare," *Journal of Military Ethics*, 3:3: 233–51.

Mander, J. (1978), *Four Arguments for the Elimination of Television* (New York: Morrow Quill Paperbacks).

Meilinger, P.S. (2001), "Precision Aerospace Power, Discrimination, and the Future of War," *Aerospace Power Journal*, 15:3: 12–20.

Metz, S. (2000), "The Next Twist of the RMA," *Parameters: US Army War College Quarterly*, Autumn: 40–53.

Murphy, J.G. (1985), "The Killing of the Innocent," in J. Sterba (ed.) *The Ethics of War and Nuclear Deterrence* (Belmont, California: Wadsworth).

Nagel, T. (1972), "War and Massacre," *Philosophy and Public Affairs*, 1: 123–44.

Office of the Secretary of Defense (2005), *Unmanned Aircraft Systems Roadmap: 2005–2030* (Washington DC: Department of Defense, United States Government).

Olsthoorn, P. (2005), "Honor as a Motive for Making Sacrifices," *Journal of Military Ethics*, 4:3: 183–97.

Peterson, G.I. (2005), "Unmanned Vehicles: Changing the Way to Look at the Battlespace," *Naval Forces*, 26:4: 29–38.

Robinson, P. (2007), "Ethics Training and Development in the Military," *Parameters: US Army War College Quarterly* (Spring): 22–36.

Scarborough, R. (2005), "Special Report: Unmanned Warfare," *Washington Times*, May 8.

Schmitt, M.N. (2005), "Precision Attack and International Humanitarian Law," *International Review of the Red Cross*, 87:859: 445–66.

Shachtman, N. (2005a), "Attack of the Drones," *Wired Magazine* 13:6, Available at http://www.wired.com/wired/archive//13.06/drones_pr.htmal, at 25.08.05.

Sharkey, N. (2009), "Death Strikes from the Sky: The Calculus of Proportionality," *IEEE Technology and Society*, 28:1: 16–19.

Silver, S.M. (2006), "Ethics and Combat: Thoughts for Small Unit Leaders," *Marine Corps Gazette*, 90:11, 76–8.

Singer, P.W. (2009), *Wired for War: The Robotics Revolution and Conflict in the 21st Century* (New York: Penguin Books).

Sparrow, R. (2009a), "Predators or Plowshares? Arms Control of Robotic Weapons," *IEEE Technology and Society*, 28:1: 25–9.

Sparrow, R. (2009b), "Building a Better WarBot: Ethical Issues in the Design of Unmanned Systems for Military Applications," *Science and Engineering Ethics*, 15:2: 169–87.

Sparrow, R. (2007a), "Killer Robots," *Journal of Applied Philosophy*, 24:1: 62–77.

Sparrow, R. (2007b), "Revolutionary and Familiar, Inevitable and Precarious: Rhetorical Contradictions in Enthusiasm for Nanotechnology," *NanoEthics*, 1:1: 57–68.

Sparrow, R. (2005), "'Hands Up Who Wants To Die?': Primoratz on Responsibility and Civilian Immunity in Wartime," *Ethical Theory and Moral Practice*, 8: 299–319.

Sparrow, R. (2004), "The Turing Triage Test," *Ethics and Information Technology*, 6:4: 203–13.

Sullivan, J.M. (2006), "Evolution or Revolution? The Rise of UAVs," *IEEE Technology and Society Magazine*, 25:3 (Fall): 43–9.

Toner, J.H. (2000), *Morals Under the Gun: The Cardinal Virtues, Military Ethics, and American Society* (Lexington: The University Press of Kentucky).

Tuttle, R. (2003), "Kill Chain Timeline Now Down To 'Single Digit Minutes'," *Aerospace Daily*, 206:58: 5.

Ulin, D.L. (2005), "When Robots Do the Killing," *Los Angeles Times*, January 30.

United States Department of the Navy (2007), *Department of the Navy Core Values Charter* 2007 [cited 17.7.2007]. Available from http://ethics.navy.mil/corevaluescharter.asp.

Walzer, M. (2000), *Just and Unjust Wars: A Moral Argument with Historical Illustrations*. 3rd edn (New York: Basic Books).

Watson, B.C.S. (1999), "The Western Ethical Tradition and the Morality of the Warrior," *Armed Forces and Society*, 26:1: 55–72.

PART IV
New Actors in the Battlefield

Chapter 8
Ethics and Mercenaries

Uwe Steinhoff

The private sale of military services to foreign parties has become a huge industry. From 1994 to 2002 the US Defense Department entered into over 3,000 contracts with military firms, with an estimated value of US $300 billion (Singer 2003; Topsy. org). Between 10,000 and 20,000 – the exact numbers are unknown – privately employed armed men are presently working in Iraq. Iraq is only exceptional with regard to the quantity. Private military firms have provided services in over 110 countries (Catan 2004; Public-i.org, 2005; Topsy.org 2005). These services include logistics, advice and even combat. Private military firms or companies (PMFs or PMCs) are working for NGOs, for the UN, for states, for rebels, for criminal organizations and for multinational corporations. Cooperation with the latter is often used to secure enormously lucrative mining licenses in weak states. Some critics have therefore spoken of an armed "corporate imperialism." Certain strong states, on the other hand, apparently host PMFs in order to promote national arms sales and to use them as proxies for pursuing certain foreign policies while simultaneously dissociating themselves from them. The PMFs and their employees, in turn, distinguish and dissociate themselves from mercenaries, presenting themselves as respectable and legitimate.

Given its extreme and growing importance, the PMF and mercenary phenomenon warrants philosophical examination in order to determine its ethical ramifications. The present chapter will focus on two questions. First, is being or acting as a mercenary per se something immoral? The answer is "no." Mercenaries, it is argued, are no worse than regular soldiers; in fact, they might even be better. However, even if a single mercenary and his mercenary activity are not necessarily immoral, the widespread use and deployment of mercenaries and mercenary organizations might still have very negative consequences. It is argued that it does indeed have such consequences, but that they are not sufficient to justify abolishing mercenaries. Since the mercenary system also has considerable ethical advantages, regulating it is preferable to abolishing it.

The Mercenary's Morality

According to the *Oxford English Dictionary* (2005), a mercenary is "a soldier paid to serve in a foreign country or other military organisation." Although, from a philosophical (and political) perspective, one can take issue with this definition

(Steinhoff 2008), it still suffices for the purposes of the following discussion. In any case, it is clearly preferable to the much too narrow (and biased) definition given by Article 47 of the 1977 Additional Protocols to the Geneva Convention.

What is so bad about a soldier paid to serve in a foreign country? Mercenaries are often reviled as "whores of war." But this is misleading. Usually a prostitute is hired for, let us say, a 30-minute or one-hour stand. One could also say that she is hired for "one round." This is *not* analogous to the mercenary. To be analogous, the mercenary would have to be hired for a specific exchange of fire: "I'll pay you *x* pounds if you sleep with me" is analogous to "I'll pay you *y* pounds if you help me against this guy over there in the bush who is shooting at me." The actual arrangements between mercenaries and clients, however, are much broader. For example, the mercenary's obligations are not restricted to firing on the enemy or protecting his client against him, or to providing military advice or training; rather, mercenaries normally also have to report from time to time to their clients. Mercenaries do not only fight for their clients, they also meet and talk with them. The mercenaries who fought variously in the Congo, Biafra and Angola in the 1960s and 1970s seem to have developed personal ties with their employers, which transformed their involvement in armed conflict into much more than a mere business venture. Besides, there was also a significant common ideological and political ground (Mockler 1986). If one is looking for an analogy to this in the world of sex, the relation between a rich sugar daddy and his mistress is a far better candidate than the customer-prostitute relation. And while prostitutes may accept more or less any customer who is able to pay, mistresses will be far more selective.

Not the least outcome of this finding is that certain criticisms of mercenarism have to be rejected. Tony Coady, for example, who makes much of the alleged analogy between prostitutes and mercenaries, endorses just war theory and claims that a soldier's "motives for fighting should at least be under the supervision of his considered belief that the war is just. Absent that belief, he will not fight: with it, he may or may not fight, depending on other motivations" (Coady 1992, 63). Based on this premise, he adduces the following argument:

> If the only legitimate justification for war is defensive, then someone who is not a citizen of the state being defended against attack may be presumed unlikely to have the sort of solidarity with and stake in the well-being of the community under threat that would provide appropriate motive for killing on its behalf. Unlikely, but not impossible. If we add that he fights principally for money, then we can practically be certain that he lacks the motive appropriate to war. We then have a ready demarcation of a group of people whose participation in war is bound to be immoral. Hence the point of our conjunctive definition [of mercenaries]. (Coady 1992, 64)

Several things have to be said here. First of all, Coady starts, it appears, by presupposing the illegitimacy of mercenarism and then proceeds to try to provide a definition that suits this presupposition. His definition, however, doesn't suit the phenomenon. The two elements of his conjunctive definition are "monetary motive and foreign affiliation" (Coady 1992, 55). But states are not the only communities one can be affiliated to, and the monetary motives of mercenaries do not have to be as strong as Coady thinks. In fact, he even grants the latter point (Coady 1992, 57ff.), though without coming up with a more appropriate definition. Such a definition would acknowledge that mercenaries may be not so much "whores of war" as "mistresses of war," and this means that they may well decide to fight only in just wars (Lynch and Walsh 2000, 141), just as a woman who wants her sugar daddy to pay her bills in exchange for sexual favors may decide to sleep only with nice sugar daddies.

Second, it is *not* clear what the point of the conjunctive definition is for Coady's moral argument. After all, the fact that someone is not a citizen of the state being defended against attack is in itself no reason at all to presume him "unlikely to have the sort of solidarity with and stake in the well-being of the community under threat that would provide appropriate motive for killing on its behalf." Obviously, an unpaid foreign volunteer is much more likely than a conscripted citizen to have the "appropriate motive." Even more important, with respect to the said likeliness there is no discernible difference between paid citizens and paid foreigners. Members of both groups can fight principally for money, and members of both groups can have their "motives for fighting ... under the supervision of [their] considered belief that the war is just." Thus, Coady fails to show that we can be "practically certain" that a mercenary "lacks the motive appropriate to war."[1]

But even if we do not have such a "practical certainty", mercenaries may statistically still be more likely to have an inappropriate motive for fighting than regular soldiers. Well, perhaps, but this is an empirical claim that does not amount to an absolute (or practically absolute) condemnation of mercenarism. Worse yet, it is not a particularly convincing claim. As Mockler dryly notes:

> The professional too the regular army officer or NCO in any army in the world – fights for money and, as a comparison between recruiting figures and wage increases show, often mainly for money and is in that sense a mercenary himself. (Mockle, 1986, 34f.)

---

1   Iain Scobbie provides an excellent criticism of Coady's use of the alleged whore/ mercenary analogy. He nicely sums it up as follows: "Coady's view of the good sex the idea of mutual enjoyment which has the character of a gift—is romanticised; he postulates an ideal case in relation to which prostitution is compared and found wanting, which simply ignores the fact that much, if not most, non-prostitutional sexual activity does not live up to this standard. Coady does not consider whether there may be something to be said for sexual congress which is simply the mutual satisfaction of lust, or for promiscuous sex as a hobby, or for recreational sex. A parallel romanticism exists in his analysis of war motives." (Scobbie 1992, 84).

Richard Holmes cites D.M. Mantell's study of American Green Berets in Vietnam
to the same effect:

> I was not talking to fervent and ideologically engaged persons but rather to
> socially and politically disinterested professional soldiers who were uniformed
> on the social and political issues of the day ... They made little attempt to
> disguise the fact that they saw themselves as hired guns, paid killers who were
> not particularly concerned with their employers or their victims. (Holmes 2004,
> 286)

Besides, if one doesn't start with a nationalist assumption, then it is anything but
clear why killing for the nation ("My country right or wrong") or for glory or
in order to make one's girlfriend or father or mother proud and happy is more
"appropriate" than killing for money. Moreover, it is much easier for a mercenary
to avoid participation in an unjust war than for a member of a regular force. A
member of the United States Armed Forces, for example, faces court martial or the
end of his career if he decides not to fight in any war that the United States engages
in. Because of this he might also not be very willing to form *considered* beliefs.
He might prefer to take his government's statements about the legitimacy of a war
at face value instead of subjecting them to closer scrutiny. A mercenary, however,
loses only one contract, that is all. This doesn't put such strong pressure on him to
turn a blind eye to the moral ramifications of the proposed contract. Moreover, a
PMF may even shun unjustified wars for business reasons – a bad reputation may
deter certain financially promising business partners, like NGOs or the UN (Singer
2004, 83).

   In sum, there is no good reason to suppose that mercenaries are more likely
than regular soldiers to fight in unjust wars. There is no good reason, either, to
suppose that their fighting in a war is driven by stronger monetary motives or
intentions than in the case of regular soldiers.

   But what about the "devotion to war for its own sake," which Mockler describes
as the "real mark of the mercenary" (Mockler 1986, 36)? Couldn't this also be a
moral mark, a mark of Cain, to put it somewhat dramatically? Singer thinks it is:

> [Mercenaries] simply harbor an open commitment to war as a professional way
> of life. That is, their occupation entails a certain devotion to war itself, in that
> their trade benefits from its existence. Although soldiers often serve to prevent
> war, mercenaries *require* wars, which necessarily involves their casting aside a
> moral attitude toward war. (Singer 2004, 41)

Physicians require diseases and benefit from them – does that mean that they cast
aside a moral attitude toward diseases?

   Let us, for the sake of argument, assume mercenaries really would like to fight
or even to kill. What would actually be immoral about this? Let us imagine a
man who enjoys killing other people. However, he respects other people's rights.

Therefore, in order to kill people without violating other persons' rights, he takes part in justified military operations – and only in justified ones. There he kills people – without exception legitimate military targets on the unjust side – out of a lust for killing, and not for the sake of the just cause. To many, such a person may because of his inclinations not appear to be particularly commendable. However, to have unattractive inclinations is not yet a moral offence. One might find a person's paedophile inclinations repulsive, but one cannot morally blame him for them. One could blame him only for illegitimately acting them out. The joyful killer of our example translates his inclination into action, but not illegitimately. He respects other people's rights. Of course, the just war criterion of "right intention," as it is traditionally understood, demands more, since it originated in the world-view of Christian theologians, and hence in a certain obsession about purity of heart rather than in concern about people's rights. This demand, however, confuses moral judgement with snooping around in other people's hearts and minds. It serves not so much the protection of liberty and individual rights as character surveillance. Its affinity to totalitarian thought-control gives us every reason to reject it.

Using Mercenaries?

So far it has been argued that it is not wrong to be a mercenary or to act as a mercenary. This does not mean that using mercenaries is morally acceptable. Conversely, even if acting as a mercenary were morally unacceptable, this would not mean that using mercenaries or allowing mercenaries to pursue their jobs would also be morally unacceptable. If it were wrong to fight and kill for money, then Executive Outcome's military campaign in Sierra Leone would have been wrong. This, however, would not automatically have made it wrong to allow EO to pursue the campaign, especially if this campaign was the only available way to save the lives of thousands of innocents from extremely brutal aggressors.

It is important to keep this in mind since certain criticisms of the use of mercenaries seem to be motivated by a visceral loathing of the mercenary. As we have seen, however, there is no basis for such loathing, at least if it is not also directed against other professional soldiers. But even if this loathing were justified, even if mercenaries were the "scum of the earth," as some depict them, it would not follow that their *activity* and *making use* of them is immoral or should, moreover, be prohibited. In order to justify such a condemnation or prohibition, independent arguments have to be provided. Seven basic arguments can be discerned in the literature:

1. The costs of using mercenaries are very high for the target nation. This is particularly so because PMCs are often paid by licenses to exploit "natural resources which might otherwise have been beneficial to the citizens of these countries, many of who [sic!] suffer due to scarcity of resources" (Olonisakin 2000, 234; see also Singer 2004, 166f.).

2. The ending of armed conflict with the help of PMCs creates only superficial peace and delays conflict resolution (Olonisakin 2000, 234).
3. Mercenaries, given that they are motivated by profit, can employ their services on the side of an immoral cause (Olonisakin 2000, 236; see also Duffield, 1998, 96).
4. Mercenaries facilitate the transfer of arms to conflict regions (Olonisakin 2000, 236).
5. The use of private armies is difficult to reconcile with the enforcement of international humanitarian law (Olonisakin 2000, 236; see also Duffield 1998, 96).
6. The beneficiaries of a profitable mercenary trade will like to continue it, and this may perpetuate conflicts and threaten political stability (Olonisakin 2000, 235f.).
7. Stability is also threatened by the availability of the often very considerable military capabilities of PMFs to many different actors, including non-state actors (Singer 2004, 169–82).

Some of these claims about mercenaries are, no doubt, correct. However, this does not yet make the case against mercenarism, for the force of these arguments against the deployment of mercenaries depends on the available alternatives. For example, yes, the financial costs of the deployment of EO have been very high for Sierra Leone, but would it have been better not to deploy EO and to let the RUF continue to slaughter civilians without impediment? And wouldn't these civilians prefer living under a superficial peace to dying in a deep-rooted ongoing armed conflict? Yes, mercenaries can fight for immoral causes, but is absolutely preventing this by abolishing mercenarism more important than defending just causes in cases where the only realistically available way to do so is mercenarism? Wasn't it better to facilitate the transfer of arms to groups opposing the RUF than to leave these groups and civilians more or less defenceless? Yes, private armies sometimes violate international humanitarian law, but isn't the risk of some such violations by EO in Sierra Leone preferable to the certainty of widespread and systematic such violations by the RUF? Is stability—which can often come in dictatorial form—always preferable to armed conflict?

In the light of these questions, simply abolishing mercenarism and leaving the rest to God doesn't seem to be an attractive option. 'Funmi Olonisakin, an abolitionist who has adduced most of the aforementioned arguments against mercenaries, puts the decisive question clearly: "If they are abolished, what will replace them? Who will handle the present crisis areas and how can the necessity for private armies be removed?" (Olonisakin 2000, 233). Her answer, which is also the answer of many other abolitionists (if we leave pacifists aside), is regional security arrangements (Olonisakin 2000, *passim*). However, there is a gap in her argument. In order to justify the abolition of mercenarism, it is not sufficient to show that it can be replaced by regional (and perhaps international) security arrangements. One would also have to show that regional or international security arrangements are *better* than mercenarism. But she shows nothing of this sort.

Let us look at this more closely. Olonisakin's prime example of such a regional conflict management is the deployment of the Economic Community of West African States Monitoring Group (ECOMOG) in Liberia. However, there are notable inconsistencies in her treatment of ECOMOG as compared with her treatment of mercenaries. For example, she claims that "ECOMOG's first enforcement operation in September 1990 achieved some positive results. Perhaps the most notable was that it halted the atrocities (although not permanently), particularly the killing of innocent civilians, which was widespread in the Liberian conflict." That is, while she does not hesitate to see halting atrocities, even if this is not done permanently, as a success in the case of the regional monitoring group, she adduces as one of "two crucial reasons" against mercenarism "that the ending of armed conflict creates only superficial peace and invariably serves to delay conflict resolution and regional stability" (Olonisakin 2000, 234).[2] But this could also be said, as she should know, about ECOMOG's operation in Liberia, where "between 1990 and 1997, the warring parties in the Liberian war multiplied and over a dozen peace agreements were broken" (Olonisakin 2000, 240). Besides, it should be mentioned that ECOMOG was not able to deploy efficiently without the help of PMFs such as PAE and ICI Oregon (Singer 2004, 182f.), and also hired single mercenaries who were looking for a new job after their former employers— PMFs such as Stabilco—had closed down or ended operations (Vines 2000, 182). In other words, ECOMOG is not a particularly good example of an *alternative* to the deployment of mercenaries. Unfortunately, this is likely to be true of practically all such forces in the African context.

Not only are the capabilities of regional forces like ECOMOG not superior to certain mercenary outfits, but the same holds for their morality and for the side effects of their deployment. ECOMOG, for example, is lead by Nigeria, the most powerful state in the Economic Community of West African States (ECOWAS). This, however, engenders serious problems, as Olonisakin admits:

> An accountable and responsible hegemony would set a good example. If the hegemony is not accountable, the opposite effect could be achieved. So far, Nigeria, though capable, has failed to instil positive standards in other member states, given the deplorable state of its own domestic environment. (Olonisakin 2000, 250)

Moreover, the problem lies not only with Nigeria:

---

2   This is also an important point for Vines (2000), who, like Olonisakin, conveniently forgets that EO had been thrown out of Sierra Leone by its then president, despite EO's warning that as a consequence fighting would continue. It did, and blaming EO for this fact seems to be not only egregiously unfair but absolutely ridiculous.

> Like the OAU [Organisation of African Unity], the Charter of ECOWAS
> contained the non-interference in internal affairs clause. It was convenient for
> the leaders of member states, which were undemocratic and had a poor human
> rights record. (Olonisakin 2000, 248)

Where, then, are the advantages of regional security forces such as ECOMOG? The
leaders of these states are no less motivated by personal profit than mercenaries.
They are mostly not acting in the interest of their peoples, let alone in the interest
of the peoples of other states (or failed states). Accordingly, African states have
"interfered extensively, but covertly, in the affairs of neighbouring states (providing
arms, rear bases and refuge for dissidents, funnelling third party weapons to
insurgents, etc.)" (Hutchful 2000, 216). It would therefore be naive to suppose
that ECOMOG has not worked, as mercenaries presumably do, as a facilitator of
the arms trade, especially given "the alliances between Taylor and certain West
African states" and "between ECOMOG and certain Liberian and Sierra Leonean
warlords" (Hutchful 2000, 222). Besides, ECOMOG needs guns for itself, too, not
only for its allies. Further, it would also be naive to suppose that Nigeria, which
"spent 30 billion Naira (US$ 3 billion) on the ECOMOG operation in Liberia"
(Olonisakin 2000, 242), did this for altruistic reasons[3] and would not be strongly
tempted to perpetuate a conflict if this were in its own financial interest (or, more
precisely, in the interest of its rulers) and helped it gain access to resources it
would otherwise not get – for example, through cooperation with the warlords in
whose region the "monitoring forces" are deployed. Another way of promoting
one's financial interests—at the cost of an only superficial peace without resolving
underlying social conflicts and injustices—lies in cooperation with, or bringing
to power of, repressive rulers in other states. Nigeria, which is "seen to export
a 'coup culture' to other countries" (Olonisakin 2000, 253), certainly has some
experience of this method. It probably goes without saying that the practitioners of
such methods will keep their rewards, in the form of power, money and resources,
to themselves instead of sharing it with their peoples; and given their poor human
rights records they can hardly be expected, in cases of armed conflicts, to be
guarantors of international humanitarian law.

Olonisakin and other contributors to the anti-mercenary volume *Mercenaries*
acknowledge these considerable shortcomings of regional security organizations.
Therefore, they demand their *improvement*:

> All the avenues through which mercenaries can be controlled in Africa have
> serious loopholes that make any such attempt a daunting task. At the heart of the
> short- to long-term strategies is the need to have clearly identifiable, responsible
> and accountable regional hegemonies. Even where they are identifiable,
> responsible and accountable, they may be incapable of coping with the sheer

---

3   We know at least that Nigeria's operation in Sierra Leone was "underpinned by the
offer of a staggering diamonds concession" (Musah 2000, 104).

magnitude of crisis. The assistance of the international community will be crucial in this regard. The nature and structure of their assistance must be seriously modified to reflect the needs on the African continent if they are to have any impact and go some way towards eliminating the need for private armies. To achieve this objective through the creation of strong regional structures is a daunting but not insurmountable task. (Olonisakin 2000, 254)

A couple of things have to be said about this. First, given the "daunting" character of the task, regional security arrangements do not at all look like a viable short-term solution. Neither is it clear that they are a viable mid-term solution. In other words, with respect to the short term and perhaps even to the medium term, the "need" for mercenaries will remain—and why shouldn't such a need be satisfied, especially if it is a need of innocent civilians otherwise likely to be slaughtered by warlords or brutal regimes? Second, the criticism of mercenaries advanced by Olonisakin and others rests on a somewhat unfair comparison: existing and *real* mercenary outfits are compared with *idealised* potential regional security arrangements.

Thus, the first six arguments given above are not sufficient to make a case against mercenaries. The seventh argument, however, has considerable force and points to much broader issues, notably to the problems connected to globalization. The problem, though, is not so much that many different actors, including sub-state actors, gain access to significant military capabilities by hiring PMFs. Sub-state actors are not necessarily worse than state actors, and whether the loss of stability is a bad thing depends on the political conditions whose stability is at issue. Hitler's regime was quite stable too (its demise came from the outside). The problem is, rather, that only those groups gain access to these capabilities *who can afford it* (Singer 2004, 187ff. and 226ff.). This can easily translate – and has already done so – into a "corporate imperialism" (Singer 2004, 188) and into the propping up of illegitimate rulers (Duffield 1998, 96) at the expense of the wider population. This is not something new. Historically, mercenaries have often functioned as palace guards for rulers with little popular support; and charter companies like the East Indian Company have used their military power to exploit other countries. History also shows that the reproach of imperialism cannot be rejected by claiming that PMFs generally help only legitimate governments, for "[s]uch a response ... misses the parallel to 19th-century imperialism, which also usually began when a weak ruler requested the original intervention" (Singer 2004, 188). (Besides, not every *de facto* ruler is legitimate.) It is true, of course, that it doesn't necessarily take mercenaries to prop up an illegitimate ruler. Many past and some present military dictatorships prove this. However, never before, largely because of the collapse of the former eastern bloc, have small guns as well as heavy weapons been as easily and cheaply available as they are now. This means that guerrilla movements in many regions of the world have much less difficulty drawing level to the national army or paramilitary groups with respect to armament than they once had. However, it is one thing *to have* a sophisticated weapon system, like, for example, a Mi-24 helicopter, and another thing to be able to operate and maintain it or to deploy one's troops in a way that makes efficient use of the weaponry. That is, military training and expertise are what is really needed.[4] But for the most part neither national warlords and guerrilla movements *nor* national armies or paramilitary forces in Africa, which is the major theatre of mercenary deployment (besides Iraq), have the required technical, tactical and strategic capabilities. Some mercenaries and PMFs do. *Their* services, however, decisive as they are, are *not* cheap. Paying for them requires not only considerable amounts of money but in some cases also costly concessions for resource extraction. This puts multinational corporations and corrupt kleptocratic leaders at a huge advantage *vis-à-vis* incipient guerrilla and resistance movements.

If the availability of PMFs with considerable military capabilities tends, as just suggested, to work under current conditions to the detriment of emancipatory movements – much more so than, for example, regional security arrangements do – is this then an argument for the abolition of mercenarism? To begin with, two meanings of "abolition" have to be distinguished: "abolition" in the legal sense and "abolition" in an empirical or actual sense. Slavery has been legally but not empirically abolished—it has been outlawed, but it still exists. Since legal prohibitions on a profitable practice can only suppress that practice, but never completely end it, it is not particularly logical to assess, as some philosophers unfortunately have done,[5] the question whether a certain practice should be legally abolished by pointing to the good its complete eradication would constitute. To apply this to our concrete issue: Even if, *ceteris paribus*, the world would be a better place if there were no mercenarism, this would not yet imply that it would be a better place if mercenarism were legally abolished. Having said this, however, we can be in no doubt that outlawing slavery has indeed made the world a better place, for it has saved uncountable numbers of people from slavery, although it hasn't saved all. Effectively outlawing mercenaries, too, would probably suppress mercenarism to such a degree that PMCs with decisive military capabilities would vanish, since such PMCs cannot exist without the permission of militarily capable domicile states.

But would this make the world a better place? After all, mercenary outfits also have advantages. For example, there are conflicts in which atrocities, even genocides, are committed, but in which those states that would be militarily capable of intervening lack the will to do so, for example because they don't want to run the risk of high casualties for their own troops. This, in fact, is typical of

---

4   The story of the RUF campaign in Sierra Leone and the later campaign of EO against the RUF is a paradigmatic example of this.

5   For example, Kant or Habermas. For a criticism of Habermas's view that every "valid norm must satisfy the condition that the consequences and side effects its *general* [by this Habermas means universal, exceptionless] observance can be anticipated to have for the satisfaction of the interests of *each* could be freely accepted by *all* affected", see Steinhoff (2009, 143–53).

many Western democracies. In such cases, a capable PMC could be the last hope and the only alternative to the continuation of the atrocities.[6]

The question is, of course, how this advantage is to be weighed against the said disadvantage. It is not an easy question. Nevertheless, answering it is somewhat facilitated by the fact that the advantage is not under all circumstances an uncompromised one. First of all, while EO in Sierra Leone, to take this famous example again, indeed halted the atrocities committed by the RUF, EO apparently also committed atrocities itself (Singer 2004, 218), though on nowhere near the same scale as the RUF. More important, however, is that the halting of atrocities is sometimes combined with the propping up or ushering into power of a repressive regime. For example, although the "international community" condemned the putsch that put president Kabbah in Sierra Leone out of office after EO had left, it should not be forgotten that Kabbah's regime, formerly propped up by EO, "was anything but democratic and peaceful" (Musah 2000, 96). On the contrary, on coming to power Kabbah pursued determinedly what William Reno (1999) calls warlord politics; that is, he downsized the national army, equipped his own private army and muzzled the opposition and independent press (Musah 2000, 95). In other words, he built the conditions on which the stereotypical repressive and kleptocratic African leader can thrive.

One might be tempted to object: Well, if that halts the atrocities, it might be worth paying the price. But this is far from certain. Sierra Leone's population is one of the poorest in the world. The average life expectancy is 37 years, and the infant mortality rate is 164 deaths per thousand births (Singer 2004, 3). These disastrous conditions are a direct effect of the kleptocratic politics that transfers the natural wealth of the country into the private hands of a few while leaving the populace to starve and to die. Although violent deaths fill the imagination much more, are much more fascinating and make much more interesting news than deaths that result from abject poverty and the concomitant lack of food, clean water and the most basic medical supplies, the latter sort of death is not better than the former (Pogge 1997). And since many times more people have been killed each year by the kleptocratic politics of Sierra Leonean leaders than by the violent campaigns of the RUF, it is arguably much better to have this politics ended by a very violent and even atrocious revolution than allowing it to continue for years, perhaps decades to come. As a German proverb goes: Better an end with terror than terror without ending.

These last considerations have dealt with the second anti-mercenary argument given above. However, as already explained, this argument applies no less to regional security arrangements. Thus, these considerations do not support the replacement of mercenaries by something else, but rather Edward N. Luttwak's suggestion: "Give war a chance" (Luttwak 1999). It is the suggestion to leave

---

6   For example, the UN at a certain point discussed the deployment of EO in the Rwandan refugee crisis in 1996. The plan was finally dismissed—to the detriment of thousands of Hutu refugees who were slaughtered (Singer 2004, 185).

internal conflict to internal parties, thereby, presumably, allowing a much more stable result to emerge. Although this will work in some cases and may be preferable to precipitate humanitarian intervention, it won't work in other cases, namely, those in which the continuation of the war is more profitable for those who wage it than a peaceful settlement, unless this settlement is itself a kleptocratic one, favoring the former war parties or the winners at, again, the expense of the wider population. This is very often the situation in resource-rich and ethnically divided countries. However, even if these latter cases were the more frequent ones, this would still not weaken the force of the above considerations since the deployment of mercenaries in such situations is likely to end the war, as in the case of Sierra Leone, by entrenching an illiberal and kleptocratic regime and hence continuing abject living—or dying—conditions for the wider population.

The result, then, of the seventh argument is that under current conditions mercenarism in underdeveloped but resource-rich countries leads to a strengthening of corporate imperialism and of illiberal and kleptocratic regimes at the expense of emancipatory movements and the wider population. This disadvantage is not outweighed by the advantage of the availability of PMCs for the halting of atrocities, since the political and social costs, especially the long-term costs, of a deployment for this purpose will in most cases be too high: that is, they will forestall the—albeit very violent—progress to better political and social conditions for the wider population. In other words, under current conditions legally abolishing mercenarism would arguably make the world a better place.

Under current conditions. Current conditions, however, include the lack of appropriate national and international regulations of mercenarism. And none of the above considerations can show that an abolition of mercenarism if preferable to its regulation. On the contrary. If effective abolition is possible, effective regulation is possible too. Such regulation of mercenarism could suppress mercenarism's disadvantages while simultaneously developing its advantages. For example, mercenary outfits could be used to halt atrocities in a way that avoids entrenching illiberal regimes. This could be done, for example, by complementing a mercenary campaign with a subsequent peacekeeping and social reconstruction mission, preferably under the leadership of the UN or a *responsible* regional organization. In such missions, mercenaries could play a role too. In fact, PMFs have already been used by the UN, and there is no reason why that couldn't be done reasonably and responsibly. Even Olonisakin admits that, as we have seen.

Besides, it should also be noted that under proper regulation the likelihood of mercenaries and PMFs waging an indiscriminate war that makes no or few distinctions between civilians and combatants is far lower than the likelihood of regular soldiers doing so. For one thing, the principal function of the armies in Africa and Latin America and, although perhaps to a lesser degree, in Asia has not been to wage war against other states but to repress their own civilian populations. Thus, the concept of "civilian immunity" is more or less alien to the tradition of these armies; hence, one should not expect them to suddenly honor the idea.

But there is also a problem with European, Australian or North American regular forces, for Western democracies generally succumb to the so-called Mogadishu factor (Robertson, 2002, 75f.): that is, Western democracies are extremely reluctant to risk high casualties for their own troops since such casualties would heavily undermine popular support for their deployment. The typical way to avoid them is to use greatly superior force and to keep one's soldiers out of the range of the enemy's weapons, for example by high-level bombing. However, such tactics make targeting and hence discrimination extremely difficult—as, for all the talk about "precision bombing," experience with the bombing of Yugoslavia, Afghanistan and Iraq demonstrates. High civilian casualties were the price for the safety of the Allied troops. Moreover, as one can learn from certain phrases often used by the US news media in coverage of the Afghanistan war, the public display of such completely asymmetric power – missiles and fighter-bombers against stones and AK-47s – can easily lead the technologically superior party to see the technologically inferior one as subhuman. This, of course, further undermines efforts to enforce the just war principles of discrimination and proportionality. A third factor to this effect is that the population of a democratic society tends to strongly identify with "their boys (and girls)" and to see their actions in a much better light than they might deserve. What would be a war crime if committed by the other side often becomes a mere "incident" if committed by one's own side. The fourth factor concerns the motivation of regular national forces. National regular troops hear a lot about the "proud tradition" and the "glory" of their armed forces, and are usually sent on missions with the idea that they are fighting for a righteous cause, perhaps even for a "holy" cause, like the "defence of the free world." They also quite often suffer from a certain nationalist fervour. All this is no help in enforcing the just war principles of discrimination and proportionality. As Lynch and Walsh (2000, 151) put it: "Societies, if not sovereign nation-states, have existed and flourished under commodified organised violence, whilst decommodified violence has been responsible for a century that will be remembered as the 'Age of Genocides'."

None of these problems occur, or they are at least considerably mitigated, with mercenaries. First, a democracy will be much more willing to risk the lives of international mercenaries than those of their own soldiers. Thus, it will not be that strongly tempted to compromise the principle of civilian immunity and proportionality for the benefit of the safety of the mercenaries. After all, "they do it for money," not for the national glory and safety, and the heightened risk is part of what they are paid for. If, then, mercenaries are required to abide by the principles of immunity and proportionality, this will also alter the choice of weapons. The power relations in battle will no longer be so asymmetrical. In addition, the public's emotional distance from the mercenaries will also lead to a more objective assessment of their actions and of the actions of their adversaries in combat. Both factors work as a restraint on the way the war is fought. This is also true of the motivation of the mercenaries. They do normally not see the equipment of their outfit as a symbol of its greater glory, and their adventurism, though less "noble," will in most cases be much healthier and more wholesome—especially for their enemies—than a violent spiritualism (or "idealism"). As Lynch and Walsh (to cite them again) aptly put it:

> It is because mercenary violence is not something engaged for deontological reasons, reasons of intrinsic value, but for material rewards, that it does not end up building a system of extermination camps, or following a MAD military strategy based on the insane thought that it is (or was) Better to be [all] Dead than [anyone] Red'. Mercenary forces do not endlessly restock and expand armaments far beyond the possibilities of use or need, and beyond the very limits of terrestrial survival. They do not require or demand shows of public adulation and respect, marching bands or Cenotaphs. In short, they do not need warfare as something *transcendent*, something fraught with possibilities of individual and communal redemption and fulfilment. Warfare is a job ... .(Lynch and Walsh 2000, 151)

In sum, from a moral perspective mercenarism has, contrary to common wisdom, certain significant advantages. It also has disadvantages. But the latter do not outweigh the former. In particular, they do not provide sufficient reasons to legally *abolish* mercenarism. They do, however, provide strong reasons for *regulation*. Mercenaries and PMCs are dangerous, and the idea, promoted by some advocates of the industry, that the "constraints" of the market and self-regulation will be sufficient to avoid undesirable developments is unbelievably naive (Singer 2004, Part III) or, more likely, intentionally misleading propaganda. For fairness' sake, though, it should be noted that not only advocates of mercenarism sometimes paint a rather one-sided picture; opponents sometimes do so too.

Bibliography

Catan, T. (2004), "Private Military Companies Seek an Image Change," *Financial Times*, December 1, 2004, retrieved from http://corpswatch.org/article.php?id=11725 on 11/2/2005.

Coady, C.A.J. (1992), "Mercenary Morality," in A.G.D. Bradney (ed.), *International Law and Armed Conflict (Archiv für Rechts- und Sozialphilosophie, Beiheft 46)* (Stuttgart: Franz Steiner Verlag), 55–69.

Duffield, M. (1998), "Post-modern Conflict: Warlords, Post-adjustment States and Private Protection," *Civil Wars*, 1:1, 65–102.

Holmes, R. (2004), *Acts of War: The Behaviour of Men in Battle* (London: Cassell).

Hutchful, E. (2000), "Understanding the African Security Crisis," in Musah and Fayemi (eds), 210–32.

Luttwak, E.N. (1999), "Give War a Chance," *Foreign Affairs*, 78:4, 45–51.

Lynch, T. and Walsh, A.J. (2000), "The Good Mercenary?," *The Journal of Political Philosophy*, 8:2, 133–53.

Mockler, A. (1986), *The New Mercenaries* (London: Corgi Books).

Musah, A.-F. (2000), "A Country Under Siege: State Decay and Corporate Military Intervention in Sierra Leone," in Musah and Fayemi (eds), 76–116.

Musah, A.-F. and Fayemi, J.'K. (eds) (2000), *Mercenaries: An African Security Dilemma* (London and Sterling: Pluto Press).

Olonisakin, 'F. (2000), "Arresting the Tide of Mercenaries: Prospects for Regional Control," in Musah and Fayemi (eds), 233–56.

Oxford English Dictionary (2005), retrieved from http://dictionary.oed.com, accessed on 1/6/2005.

Pogge, T. (1997), "The Bounds of Nationalism," *Rechtsphilosophische Hefte 7*, 55–90.

Public-i.org (2005), "Making a Killing: The Business of War," retrieved from http://www.public-i.org/report.aspx?aid=177&sid=100, accessed on 11 February.

Reno, W. (1999), *Warlord Politics and African States* (Boulder, CO and London: Lynne Rienner).

Robertson, G. (2002), *Crimes Against Humanity: The Struggle for Global Justice* (London and New York: Penguin).

Scobbie, I. (1992), "Mercenary Morality: A Reply to Professor Coady," in A.G.D. Bradney (ed.), *International Law and Armed Conflict* (*Archiv für Rechts- und Sozialphilosophie, Beiheft 46*) (Stuttgart: Franz Steiner Verlag), 71–91.

Singer, P.W. (2003), "Peacekeepers, Inc.," *Policy Review*, 119:3, 59–70, retrieved from http://proquest.umi.com.

Singer, P.W. (2004), *Corporate Warriors: The Rise of the Privatized Military Industry* (Ithaca, NY and London: Cornell University Press).

Steinhoff, U. (2008), "What Are Mercenaries?," in Andrew Alexandra, Deane-Peter Baker and Marina Caparini (eds), *Private Military and Security Companies: Ethics, Policies and Civil-Military Relations* (London: Routledge), 19–29.

Steinhoff, U. (2009), *The Philosophy of Jürgen Habermas: A Critical Introduction* (Oxford: Oxford University Press).

Topsy.org (2005), "Military Contractors," retrieved from http://topsy.org/contractors.html, accessed on 11 February 2005.

Vines, A. (2000) "Mercenaries, Human Rights and Legality," in Musah and Fayemi (eds), 169–97.

Chapter 9

Ethics and the Human Terrain: The Ethics of Military Anthropology

George R. Lucas, Jr.

The new era of "irregular warfare" (IW) is thought to consist primarily of the substantial changes and challenges in military tactics and military technology – the so-called "revolution in military affairs" (RMA). On the one hand, asymmetric warfare, including a variety of types of terrorist attacks involving rudimentary technology, has largely replaced the massive confrontations between opposing uniformed combat troops employing high-tech and heavy-platform weaponry that previously characterized conventional war. On the other hand, new kinds of "emerging technologies," including military robotics and unmanned weapons platforms, non-lethal weapons, and cyber-attacks, ever-increasingly constitute the tactical countermeasures adopted by established nations in response.

The era of irregular or unconventional war also, however, requires another revolution—a cultural sea-change that has been much harder to identify, and much slower in coming. This radical cultural shift entails re-conceptualizing the nature and purpose of war-fighting, and of the war-fighter. This relatively neglected dimension of RMA requires new thinking about the role of military force in international relations, as well as about how nations raise, equip, train, and, most especially, about the ends for which these nations ultimately deploy, their military forces. The new era of IW requires that military personnel develop a radically-altered vision of their own roles as warriors and peace-keepers, as well as appropriate recognition of the new demands IW imposes upon the requisite expertise, knowledge, and limits of acceptable conduct for members of the profession of arms. Finally, the new era of IW requires that those institutions and personnel who educate and train future warriors be much more effective in helping them understand, and come to terms with, their new identity, the new roles they will be expected to play in the international arena, and the new canons of professional conduct appropriate to those roles.

In the current Afghan campaign, as well as in the war of intervention in Iraq, this goal was addressed in what turned out to be a rather controversial manner: namely, through the extensive use of scholars and academics in key advisory roles in combat zones. Here I am not speaking of the longstanding reliance, back at the home front, on engineers and technical specialists in weapons development, nor the use of political scientists, economists, and international relations experts from universities and think tanks to serve as advisors on DOD or State Department policy committees. Those academic experts have always been ready, from the comparative safety of their domestic-based policy circles, to proffer advice to American political leaders and policy makers on where, when, and against whom to risk other peoples' lives and welfare in going to war against America's perceived adversaries (what Cheney Ryan, in his new book, denounces as "alienated war").[1]

Instead, the "new wrinkle" in this respect is the use of academic specialists in cultural and regional knowledge – in language, history, archaeology, and anthropology – both to prepare American combat troops for deployment in far-flung and unfamiliar regions of the globe – and also, in an unprecedented fashion, *to train for deployment themselves*. Rather than simply educating others, academic scholars and subject matter specialists are being invited directly to serve in uniform alongside military troops, advising brigade combat teams (BCTs) and provisional reconstruction teams (PRTs) on local culture and customs. Anthropologists, in particular, long consigned to the isolated study of what is often disparagingly termed "the exotic and the useless," have recently found themselves being courted in unprecedented fashion by the U.S. Department of Defense and paid rather handsomely (upwards of $12–20,000 per month) to serve in a variety of roles in the current campaigns in Iraq, and especially in Afghanistan. Marine Corps General James H. Mattis, commander of the United States Central Command (Tampa, FL) is credited with the observation that, in this new era of unconventional or "irregular" warfare – counterinsurgency, humanitarian intervention, peace-keeping and stability operations – our ground forces need to be able to "... navigate cultural and *human terrain* as easily as Marines can now use a map to navigate *physical terrain*."

The concomitant rise of a relatively new field, dubbed "military anthropology" (and by its severest critics, "mercenary anthropology") has provoked a firestorm of controversy among academic anthropologists, particularly among the membership of our sister society, the American Anthropological Association (AAA). Anthropologists express concern over the possible misuse or distortion of their disciplinary expertise, and even more over prospective violations of their discipline's professional Code of Ethics[2] that might transpire during the course of these activities. The present controversy also harks back to what many regard as the unsavory involvement of anthropologists, sociologists, and other social scientists in the infamous American-led counterinsurgency campaigns of the Cold War-era in Latin America and southeast Asia during the 1960s.

The principal moral debate properly centers on the US Army's "human terrain system" (HTS), a pet project of US Army General David H. Petraeus, Commander

---

1   Cheney Ryan, *The Chickenhawk Syndrome* (Lanham, MD: Rowman & Littlefield, 2009).

2   Adopted in 1998, and extensively revised in light of the present controversy in February, 2009, the AAA Code of Ethics may be found on the Association's website at: http://www.aaanet.org/issues/policy-advocacy/Code-of-Ethics.cfm.

of the US Central Command, and former Commanding General of Multi-National Forces in Iraq. Petraeus, who holds a doctoral degree from Princeton University in international relations, has been especially solicitous and respectful of academic expertise in his conduct of complex counterinsurgency campaigns, and has found a strong supporter in this newfound military respect for academic erudition in the current Secretary of Defense, Robert Gates, who is himself a former university president with a Ph.D. in history. Thus, on October 31, 2007, the Executive Board of the AAA issued a statement strongly opposing HTS, and strenuously objecting to the participation of members of the Society in this program:

> In the context of a war that is widely recognized as a denial of human rights and based on faulty intelligence and undemocratic principles, the Executive Board sees the HTS project as a problematic application of anthropological expertise, most specifically on ethical grounds. We have grave concerns about the involvement of anthropological knowledge and skill in the HTS project. The Executive Board *views the HTS project as an unacceptable application of anthropological expertise.* (AAA 2007)[3]

Those preliminary conclusions were reaffirmed in December 2009, when the AAA's Commission on the Engagement of Anthropology with US Security and Intelligence Community (CEAUSSIC) released its second, in-depth study of the HTS program, suggesting "that the AAA emphasize the incompatibility of HTS with *disciplinary ethics and practice.*"[4]

One might well wonder how, in point of fact, this new approach to the use of academics in irregular warfare is itself faring. On October 5, 2008, an anthropologist working on a Human Terrain System team in Iraq posted a "blog" of his impressions after two months in the field. "The best description," he wrote,

> ... is that it's like, well everything in life. I get excited about the work, I get discouraged. I feel like I am doing things that can have long term value and I wonder what the hell I'm doing in this screwed up place. I have learned that the backs of my ears may never be clean again, the ex-pat life agrees with me, I miss beer and sushi, and now I know what it feels like to take pictures of young men that die two days later.

---

3   The AAA website describes the Executive Board's positions in varying terms, from "opposing" to "strongly disapproving" of HTS as a violation of the AAA Code of Ethics.

4   See http://dev.aaanet.org/issues/policy-advocacy/CEAUSSIC-Releases-Final-Report-on-Army-HTS-Program.cfm. See also the coverage of these developments in the *Chronicle of Higher Education* by David Glenn, "Program to Embed Anthropologists with Military Lacks Ethical Standards" (December 3, 2009): http://chronicle.com/article/Program-to-Embed/49344/.

Depending on the day you ask me, I will say that the problems with the entire HTS program are so insurmountable it should be started over from scratch, and on others, I see progress. We focus on being as much help as we can to our brigade in their efforts to help improve living and security conditions for local people as best we can.. I go from wanting to quit to wanting to stay here for at least a year because there always seem to be another interesting project we can do.

In other words, it's a job just like jobs everywhere.

Well, not entirely. Shortly after, in May 2008, social scientist and HTS team-member Michael Bhatia, working in eastern Afghanistan, was killed in an IED roadside attack, for which the Taliban subsequently claimed credit. Then, on November 4, 2008, Paula Loyd, assigned to US Army team AF-4 Blue, was conducting interviews among the local population in the small village of Chehel Gazi in southern Afghanistan. According to witnesses, she approached a man carrying a fuel jug and they began discussing the price of gasoline. Suddenly the man, Abdul Salam, doused her with the fuel in his jug and set her on fire. She suffered second-and third-degree burns over 60 percent of her body.[5] Tragically, Loyd died of her injuries a few weeks later in early January, 2009. One of her HTS teammates, Don Ayala, subsequently apprehended, shot, and killed her assailant, and was subsequently charged with second-degree murder by the US Army Criminal Investigation Division under the Military Extraterritorial Jurisdiction Act (MEJA), and eventually convicted of manslaughter in the US District Court of Eastern Virginia. Setting women on fire is often inflicted as a punishment for immodesty. Immodesty, however, was probably not the crime of a third HTS social scientist, Nicole Suveges, killed in Iraq in July, 2008, along with eleven others, when a bomb detonated inside the Sadr City District Council building in Baghdad. These deaths and injuries top the list of grievances of critics of the program.

Confronting the Controversy over Military Anthropology

This history, and the moral perplexities inherent in the present controversy over military anthropology, are the topic of my own recent book, *Anthropologists in Arms*.[6] What makes this controversy interesting for philosophers and indeed, for the public at large, is the manner in which an admittedly complex problem of professional ethics, defining the limits of acceptable professional conduct,

---

5   See Schachtman, (2008a), Constable, (2009), and Stockman, (2009). See also "Rough Terrain" by Vanessa M. Gezari, *The Washington Post Magazine* (August 30, 2009), pp. 16–29.

6   *Anthropologists in Arms: the Ethics of Military Anthropology* (Lanham MD: AltaMira Press, 2009).

is intricately interwoven with a wider moral debate over the justification of America's "irregular" wars in Iraq and Afghanistan. In what sense and in what way do, or ought, widespread doubts about the legitimacy of armed conflict constrain the behavior of professionals—physicians, psychologists, clergy, and perhaps anthropologists—asked to participate in some fashion in that conflict? Philosophers, in particular, might see it as their contribution to be able to bring some order and structure to this otherwise inchoate debate, perhaps by trying to distinguish, for example, between the broad public discussion of the moral justification of war on one hand, and the more narrowly focused discussion of the specific responsibilities of professionals on the other. Moral philosophers might also assist by providing a common language and conceptual framework (including a body of thoroughly-examined moral theories) that might serve to disentangle and clarify the various threads of this controversy. Those were certainly among the objectives I pursued in my recent book on this controversy.

As noted, the wider debate has been about "military anthropology," encompassing a variety of distinct activities, including, but not limited to the HTS program's dramatic efforts to "embed" anthropologists with military troops in combat zones (in Afghanistan, Iraq, East Timor, and other locations), in order to assist military personnel engaged in irregular warfare with advice and consultation regarding strategic features of the local and regional culture.

Even HTS itself, however, let alone the wider framework of "military anthropology" more generally, includes a great deal more than this specific and controversial "human terrain team" deployment program (see Table 9.1).[7] The larger HTS project also encompasses somewhat less controversial efforts undertaken by anthropologists and other social scientists to provide advice, expertise, and the results of anthropological research on "culture," and on the details of specific cultures, to military organizations for more general guidance in the formulation of effective strategy and tactics in war zones. Thus anthropologists at the Marine Corps "Center for Advanced Operational and Cultural Learning" (CAOCL) at the Marine Corps University in Quantico, VA, have aided the Marine Corps in composing new handbook for operational culture (Salmoni and Holmes-Eber, 2008).[8] Anthropologists have likewise been employed by the government to

---

7 For a wide-ranging survey of the components of this field, see the essays by a number of leading military anthropologists in *Anthropology and the United States Military: Coming of Age in the Twenty-First Century* (Frese and Harrell, 2003). For a critical account from a military practitioner perspective of how the human terrain teams approach is both failing, and undermining the military's own efforts to enhance cross-cultural competence, see "All our Eggs in a Broken Basket: How the Human Terrain System is Undermining Sustainable Military Cultural Competence," by Major Ben Connable, US Marine Corps: *Military Review* (March–April 2009), 57–64.

8 Anthropologists Barak A. Salmoni (Ph.D., Harvard University) and Paula Holmes-Eber (Ph.D., Northwestern University) co-authored this new handbook, *Operational Culture for the Warfighter: Principles and Applications* at CAOCL.

Table 9.1   Types of Military Anthropology

Symbol	Type	Description
MA_1	Anthropology of the Military	Anthropological Study of Military Culture
MA_2	Anthropology for the Military	Human Terrain Systems (HTS)
MA_3	Anthropology for the Military	educational programs (language, culture, regional studies) at military academies

write guides and study materials on local cultures for military personnel deployed around the world. With the assistance of anthropologists, for example, the cultural programming unit of the Marine Corps Intelligence Activity (MCIA), also located in Quantico, Virginia, has produced a series of training and educational materials for its troops stationed overseas, including so-called "smart cards" that summarize the "most essential features" of cultures encountered in nations as diverse as Chad, Sudan, and the Philippines, as well as in Iraq and Afghanistan. Finally anthropologists engaged in the broader HTS project have assisted the US Army in composing two new field manuals: FM 3-24, on "Counterinsurgency Warfare" or COIN (Petraeus, 2007), and more recently, FM3-07 "Peacekeeping and Stability Operations" (October 2008). The aim of all these learning aids is to provide a rapid and readily-available orientation to locale for young men and women of high school age and education who may never before have traveled far from home, let alone resided or worked in some of the exotic and unfamiliar locations to which such individuals now find themselves routinely deployed in this era of irregular warfare.

While all of the foregoing HTS activities constitute important forms of military anthropology, the latter, much broader term, also encompasses the employment, by the US military services, of anthropologists who perform routine educational and scholarly tasks for military and State Department personnel. Anthropologists teach and carry out their own individual scholarly research at federal service academies, war colleges, and language institutes. Anthropologists advise their academic employers in these institutions on how to increase cultural literacy, promote and enhance foreign language acquisition and competence among their students, and increase the "cultural awareness" and cultural sensitivities of those students. Anthropologists are being asked to assist in the development of new "regional studies" programs for the Department of Defense and its constituent military organizations. Under the code name "Project Minerva," Secretary Gates recently sought to encourage, and to generously fund, broad-based scholarly contributions to national security studies from sectors of the academic and higher education community (including the discipline of anthropology) that have heretofore been under-represented and marginally utilized for such purposes.[9]

Finally, the term "military anthropology" can be applied to a series of activities seemly distinct from all those preceding; namely, making the military itself, or its distinct organizations and/or service sub-cultures, the *objects* of anthropological study and field research. In this third, distinct sense, the military anthropologist does not render some autonomous culture or society the object of investigation in behalf of purposes entertained *by* the military. Rather, the anthropologist renders the members and sub-cultures of the military *themselves* the objects of ethnographic study. The purpose here is first and foremost simply to understand those organizations and sub-cultures more completely, as objects of scientific study, much as one is curious about the members of an alien or radically unfamiliar culture one might encounter. The results of such study might simply satisfy scientific curiosity, help the military services better understand (and perhaps improve) their own organizations, or even help societies better understand the nature and role of the military organizations with whom they co-exist.

At one extreme of this debate are anthropologists involved in the most controversial aspects of this new venture, advocating greater involvement of their colleagues in efforts to save innocent lives, reduce troop casualties, and aid in the successful rebuilding of devastated civic infrastructures, especially in Afghanistan and Iraq, two nations ravaged by decades of virtually continuous warfare and civil strife (Kilcullen, 2009; McFate 2005a, 2005b). At the other extreme are critics of any involvement of behavioral scientists and scholars with the government and military, who denounce initiatives like the human terrain teams as "mercenary anthropology," or the "militarization" of anthropology (Gonzalez 2007, Gusterson 2007, NCA, 2007). Caught in the cross-fire between proponents of HTS and anti-war activists are the bulk of those who would identify themselves as "military anthropologists" more broadly, who complain that their work is unfairly caught up in, and their own efforts and careers unfairly impugned by, this raging controversy over merely one, controversial program in a much wider and, for the most part, morally benign area of inquiry.

Because of these inflammatory concerns over the HTS program specifically, many anthropologists have gone so far as to denounce *any* cooperation with the government or military *whatsoever* as "unethical," including even some of the more apparently benign scholarly studies or educational activities described above. Professor Terrence Turner of Cornell University, a persistent critic of both military anthropology and, more broadly, of forms of practical, applied, or "practice" anthropology in non-academic settings, for example, firmly believes

---

9 Criticisms and concerns from anthropologists regarding the orientation, administration, and likely impact of the Minerva program can be found in an AAA press release dated July 7, 2008, and located on the Association's website at URL: http://www. aaanet.org/issues/press/upload/Advisory-Anthropologists-Critique-Pentagon-s-Minerva. pdf.

that "classified work for the military is unethical ... and the association should have the will and guts to say so" (Jaschik, 2009).

From Principles to Practices

My own analysis of the ethical controversy over military anthropology (Lucas, 2009), however, highlighted a glaring defect in the entire discussion of military anthropology to date. Thus far, that public debate about the moral legitimacy and professional propriety of HTS has been *a debate about principles* – either disagreements about what moral principles might be placed in jeopardy by HTS, or about basic canons of professional practice that might be found at odds with the demands placed upon the likely activities of HTS teams. It has not, however, been a debate grounded in specific evidence, or widespread experience.

When we move from an abstract debate of ethical principles to a discussion of the morality of concrete practice, for example, we might wonder more about the moral dilemma of putting lightly-trained social scientists (along with embedded journalists) at risk in combat zones. Do they constitute a danger to themselves, or to military personnel assigned to protect them, for example? Are the risks to themselves and the liability to others that they represent by their presence worth the benefits they might provide to those same military personnel in carrying out their stability and peace-keeping operations?

Rarely, in any profession or discipline, are abstract discussions of principles, or even past history alone allowed to settle, and particularly to foreclose *a priori* an area of activity that might otherwise hold promise, especially if the nature of that promise, as formally presented, was to aid and protect vulnerable victims of war. Instead, as in most disciplines, professions, and vocations, the genuine moral challenges are to be found arising in specific circumstances, in that crucible of issues that constitute case literature grounded in actual professional experience.

Hence, in my book, I attempted to sort out some of the *different types* of HTS activities, and then proceed to evaluate the ethical compliance and moral probity of each (see Table 9.2). First, there is what I labeled HTS_1, the activity of embedded field anthropologists directly engaging in strict cross-cultural translation in support of combat and security operations with war zones. Team members talk to villagers and then pass along their interpretation of what villagers say to members of the military. What they learn is for immediate and local consumption, guiding the actions and reactions of civil and military forces in zones of conflict.[10] This is what the "embedded" military anthropologist, Mark Dawson, quoted at the beginning of this paper, describes himself primarily as doing as part of an HTS team. It seems

---

10   Here I am indebted to Anna J. Simons for originally suggesting this line of inquiry in her review and critique of my earlier article in the *Journal of Military Ethics* (Lucas, 2008). This, and some of the remaining descriptions of specific activities, follow from her suggestions.

Table 9.2   Forms of HTS Activities

Symbol	Description
HTS$_1$	Providing cultural advice and regional knowledge (including language skills) on site to military personnel in combat zones
HTS$_2$	Populating non-classified, nonproprietary cultural databases maintained in the US
HTS$_3$	Cultural espionage; gathering clandestine cultural data for classified databases ("Thailand Affair")
HTS$_4$	Forensic anthropology; investigation of possible war crimes
HTS$_5$	preservation of valuable cultural patrimony in war zones

pretty clear that this is the activity that Paula Loyd was engaged in when she was brutally attacked. From all the limited accounts to date, HTS$_1$, appears to comprise the bulk of HTS activities carried out by anthropologists who are deployed and embedded in combat zones.

On a live radio interview in October, 2007, for example, Lt. Col. Edward Villacres, a Middle East historian from West Point deployed at the time to command an HTS team in Iraq, described to anthropologist David Price and HTS program director Montgomery McFate how his four-person team (consisting of an adjunct professor of anthropology from Cal State-East Bay, a junior officer fluent in Arabic, and a civil affairs specialist who is an Army staff sergeant) attempted to "enhance the ability of ground forces to make good decisions" (Rehm, 2007). He recounted how rival tribal leaders in war-torn areas have a specific cultural "formula" for reconciliation negotiations following a conflict. It is imperative to be aware of, and to subscribe to the rituals of this formula in order to conduct effective negotiations, and help the warring factions make peace. It was the HTS team's job to interpret these vital cultural data to the American troops engaged in peace-making in contested areas of the "Sunni triangle" in Iraq.[11]

Not coincidentally, this is the kind of HTS activity regarded with the greatest suspicion by critics of military anthropology as fundamentally incompatible with the AAA Code of Ethics (CoE, 1998; rev. 2009). Anthropologist David Price, on the same program, reiterated the results of his own historical research on the military's use and abuse of social scientists during World War II and the Cold War. He complained about one Pentagon official who had compared HTS favorably with the "civil operations revolutionary development support" (CORDS)

---

11   "Anthropologists and War" (Rehm 2007). Guests included McFate, Price (via telephone from the West coast), Lt.Col. Villacres (via telephone from Iraq), Col. John Agoglia, director of the US Army's "Peacekeeping and Stability Operations Institute," and David Rohde, a renowned reporter from the *New York Times,* whose article on HTS teams from the previous week is cited below.

programs undertaken during the Vietnam war.[12] That effort, Price explained, helped identify Vietnamese communist insurgents and Vietcong collaborators. It entailed Army "Green Berets" (Price alleged) "illegally" translating into English a detailed ethnography of Highlanders in Vietnam by a French anthropologist, Georges Condominas,[13] and then using the insights gained from this work to target and assassinate Vietcong collaborators. He acknowledged that he himself was not charging that such things are taking place now in Iraq and Afghanistan, but he understandably found it "troubling that a Pentagon spokesman would make such a comparison."

The foregoing discussion, however, suggests a further distinction, HTS_2, in which the anthropologist gathers information to populate databases located back in the US, accessible to analysts who are not anthropologists. These databases are proprietary to the military, but not classified, so that results can be accessed and used by other anthropologists engaged in legitimate scientific work. In her account during this radio interview, anthropologist Montgomery McFate described how she had originally proposed something like this as the main activity of HTS when she was first approached to undertake this new task in 2003. The Army, however, wanted something different, what she described as "an angel on the shoulder," that is, HTS_1, mediated by a real person, and not merely a database, and "there" (meaning, in the field) rather than "here" (in the US).

Academic anthropologists and critics of HTS often voice a concern about undertaking "clandestine research," or, as Professor Price does, fret over the possible misuse and misappropriation, without knowledge or permission, of prior anthropological research (such as the published field notes of Georges Condominas). That, however, would constitute something different from HTS_2, something more akin to genuine espionage or "intelligence gathering" than legitimate anthropological research. Accordingly, I label this very specific and focused concern HTS_3, in which the primary activity of the anthropologist (or those purloining the results of anthropological studies) is to gather information to populate classified databases. As illustrated in Price's comments, these concerns over anthropologists engaging in espionage and clandestine research activities constitutes a holdover from what anthropologists consider their "darkest days" during Vietnam, specifically, an Army project known at the time as "Project

---

12   In Price and Gonzalez, (2007): 3, this remark is attributed to David Kipp of the Foreign Military Studies Office at Fort Leavenworth, KS, who reportedly described HTS as "a CORDS for the 21st century."

13   The reference is to Condominas (1994). It is not clear why translating a published work into the readers' language is "illegal" unless the intent was to publish in violation of copyright. The point is, rather, that research results were being appropriated and used contrary to professional intent and the constraints of professional ethics.

Camelot," and a subsequent DARPA-sponsored project in southeast Asia known widely as "the Thailand controversy."[14]

This third category appears to encompass the recent allegations of the Zapotec in Mexico, who allege that a geographer and anthropologist from the University of Kansas solicited their consent, under false and deceptive circumstances, to engage in an ethnographic mapping of indigenous communities. They claim that, unknown to them at the time, his research was partially funded by the Foreign Military Studies Office of the US Army, which maintains a proprietary global database used in the Human Terrain Systems project.[15] As I will demonstrate below, however, there are no reports of job openings for, or other reports apart from this allegation of, anthropologists engaging in HTS_3, which would constitute the most explicitly problematic of these three types of activities from the standpoint of professional ethics. HTS anthropologist Mark Dawson does not acknowledge engaging in such activities. When they were injured or killed, neither Bhatia, Suveges, nor Loyd were engaged in such clandestine activities. No published or broadcast accounts either document instances, or even suggest that such clandestine activities are part of this program. At best, as seems to be the case with Price, this concern constitutes a kind of historical anxiety through which lens one refracts the prospects for commission of such abuses in the present.

In fact, however, the preponderance of work to be found, apart from working on education, doctrine and training back home, is for HTS_1. It is possible, but not confirmed, that some of the data and experiences gathered will find its way into proprietary but unclassified data bases back home. HTS_2 does have a certain allure in holding out to middle east regional experts the prospect of regaining access to regions of central Asia, such as Afghanistan, long off-limits to them as zones of conflict. But populating databases is not the principal motivation for the HTS program itself. Rather, as we will note in job ads and position descriptions, the prospect for ethnographic field research and data-gathering is part of the wider lure to research anthropologists, a potential "added personal benefit" of their HTS employment that (the job openings promise) will allow them to pursue their own scholarly research and publication on the people and "cultures" in which they are immersed during their deployment, should they wish to do so. That is what the job

---

14 See my account of these and other controversial engagements by anthropologists with the military in *Anthropologists in Arms*, Chs. 1–2 (Lucas, 2009). See also Silvert, (1965); Horowitz, (1967); Fluehr-Lobban, (2003a), and, for the "Thailand Affair" specifically, see Berreman, (2003); Price, (2003).

15 It is unclear whether their objection is to the failure of full disclosure, including funding sources and purposes to which the ethnographic data will be put, or obtaining "informed consent" allegedly under false pretenses, or, as with David Price, fear of the eventual use to which such data might be put by researchers. See "Zapotec Indigenous People in Mexico Demand Transparency from US Scholar," by Saulo Araujo of the Union of Organizations of the Sierra Juarez of Oaxaca (UNOSJO), reported January 22, 2009 (http://elenemigocomun.net/2059). There has as yet been no verification that this research, or the personnel conducting it were, in fact, connected to the HTS program.

ads promise enticingly, and such a benefit is far from constituting the sinister or clandestine activity defined as HTS_3.[16]

Fact versus Hypothesis

In my book, I consider a hypothetical case in which a democratic nation's government has embarked upon a preventive war of dubious justification, gotten itself into quite a mess, and now realizes it needs to change course, patch up the damage, and leave (Lucas 2009, Chapter 4; see also Lucas 2008). Its military now finds itself enmeshed in a nasty and seemingly-intractable counterinsurgency, and reaches out to anthropologists to assist with greater understanding of the regional culture in order to provide effective and workable solutions toward political stability, get the local populace working together to rebuild their civil infrastructure, reintroduce reasonable security under the rule of law, and then pack up and "go home."

In the book, I conclude that, in such an instance, our hypothetical anthropologist would not automatically be forbidden, ethically or morally, from providing such assistance. Specifically, providing such assistance would not constitute a violation of the AAA Code of Ethics (CoE, 1998; rev. 2009), provided the motivations for this engagement were as described, and the constraints on allowable activities that these motivations imposed were indeed met. What is problematic about this hypothetical case is not the conclusions (even if some are displeased with them), but the lack of material and cultural substance in the example.

"The military," for example, is not a single, monolithic entity that entertains objectives, or even goes out and hires employees. There are competing military services, with complex internal service cultures and subcultures, and conflicting chains-of-command. "The military's objectives," like those of a nation, may reflect the fundamental proposals of some governing elite, but that does not automatically

---

16   The initial report on military anthropology released in November 2007 by the AAA's Ad Hoc Commission on the Engagement of Anthropologists with the US Security and Intelligence Community (CEAUSSIC) suggested two other kinds of HTS embedded activities: HTS_4 (providing forensic advice or examining evidence of combat casualties in pursuit of war crimes), and HTS_5 (lobbying, and indeed, actively working for the preservation of valuable cultural patrimony that happens to be located in war zones). The former would involve the forensic anthropologist in a perfectly legitimate form of law enforcement, including the possible investigation and prosecution of war crimes against civilian noncombatants. I see no prospect for this constituting a violation of any professional norms. Likewise HTS_5 (assuming the good faith and integrity of anthropologists on the project) is not a moral "problem" but a moral *obligation*: anthropologists (including archaeologists) are enjoined by the very nature of their discipline to refrain from damage or theft themselves, and to do all in their power to resist damage to, or theft of valuable cultural patrimony, sacred to the people and cultures they study, and vital to the advancement of cultural knowledge worldwide.

translate into compliance up and down the chain of command, even granting basic "buy-in" and good faith in carrying out those objectives. There are political factions within each service, and within the Department of Defense, vying for power and influence.

Typically an influential member of the governing elite, a figure like General David Petraeus in the Army, for example, formulates the idea that "the military mission" in a nation like Iraq or Afghanistan (however that mission itself is, in fact, generally characterized) might be more effectively accomplished with the help of anthropologists, who might bring an enhanced understanding the "cultural terrain." This proposal automatically becomes the brainchild or "pet project" of one faction of that branch of the military services, perhaps shared or agreed to by others, and largely ignored, or even actively opposed by yet others as a "waste of time and money," or unworkable.

Even apart from such political fragmentation, there are practical questions that bear heavily on the actual practice proposed as a remedy. For example, who will recruit and hire these scholars? How will they be oriented to the rigors of this particular kind of field work (largely unlike any most will have encountered before)? How will they be organized, deployed "within country," commanded, even evaluated? To whom will they report, and to what uses will their information finally be put in the field, far from the headquarters of those leaders whose "brainchild" their inclusion in these delicate, fragile, and (as we now see) risk-fraught military operations was in the first place?

Given all the assumptions packed into the above-mentioned hypothetical case, we would need some rather robust assurances that practical and procedural questions like these would admit of some decidedly positive answers, in order to address the manifold and substantive concerns of critics of this program. Instead, the scanty evidence we have in fact ranges from random and anecdotal testimony (such as the remarks of HTS anthropologist Mark Dawson, cited at the beginning of this essay), that the record of accomplishment is decidedly mixed (with some positive results, and no dilemmas of "professional ethics"), through rosy accounts of success from the Secretary of Defense, all the way to increasingly exaggerated accounts of death and injury, and journalistic (but unsubstantiated) charges that the actual, functioning HTS program is, on balance, an unmitigated disaster, "in total disarray," and largely amounts to a useless waste of time and resources.

Just as importantly, *we would want some kind of external oversight and evaluation of these ongoing arrangements*, so as not to have to rely solely on the reports of the individuals hired, or heads of organizations hiring them, let alone of journalists and bloggers transparently fronting for critics of the program.[17] All

---

17   One such persistent critic is Maximilian ("Max") Forte, a professor of anthropology at Concordia University (Montreal, Canada), who routinely posts articles with blog commentary and readers' replies, on his website, "Open Anthropology." He describes himself as relentlessly opposed to "institutional and disciplinary anthropology, insofar as it has or may continue to support, justify, participate in, or abide by imperial projects."

these are now involved in an inherent conflict of interest that might cloud their own evaluations of the successes or failures, as well as of the professional probity and moral rectitude, of their activities. We would need, as military officials themselves often say, "to trust, but verify."

Up until January 2009, the principal responsibility for recruiting, hiring, training, compensating, deploying, and ultimately evaluating "social scientists" hired to serve in various components of the Human Terrain System project was delegated via competitive bidding to a private firm, BAE Systems, Inc.[18]

The job descriptions below were posted in the recent past on BAE's non-classified, public web site, soliciting anthropologists to work on the Human Terrain System. Positions include openings for social scientists, research managers, HTS "analysts" and "team leaders.[19] The positions include an initial training period, typically of four months duration at the company's facility in Ft. Leavenworth, Kansas, "including orientation to the military/deployment environment, in-depth country briefings, and multi-disciplinary social science concepts and methods." Training and orientation are followed by deployment to Iraq and/or Afghanistan, typically for six to nine months. This is how a prospective applicant would initially be introduced to the program as an opportunity for his or her prospective career:

> The Human Terrain System (HTS) is a new Army program designed to improve the military's ability to understand the local socio-cultural environment in Iraq and Afghanistan. Human Terrain Teams help military commanders reduce the amount of lethal force used, with a corresponding reduction in military and civilian casualties, allowing the commander to make decisions that will increase the security of the area, allow other organizations (local and international) to more effectively provide aid and restore the infrastructure, ensure that US efforts are culturally sensitive, promote economic development, and help the local population more effectively communicate their needs to US and Coalition forces. HTS teams act as advisers to Army Brigades and Marine Corps Regiments. [An HTS Team] will not engage in combat missions, nor does it collect intelligence. [An HTS] team will [typically] analyze data from a variety of sources operating

His perspectives on the discipline and its habits, as well as specific comments on the HTS project, can be found at: http://www.openanthropology.wordpress.com.

18　In January, 2009, partly as a response to the problems described below, the BAE Systems contract was terminated, and the HTS program directly subsumed under TRADOC, the US Army's training and doctrine command. The new regime may be studied at their website, established January 15 2009, at http://humanterrainsystem.army.mil. The job advertisements and position descriptions detailed below (current as of November 30, 2008) have been largely transposed intact to this new site. The positions are now, however, standard civilian DOD positions in the US Civil Service, rather than PMC positions. With those changes, most of the remainder of the present analysis remains intact.

19　These position descriptions have been migrated largely intact, and are now posted at the HTS TRADOC website: http://humanterrainsystem.army.mil/employment.html.

in theatre (e.g. conventional military patrols, non-governmental organizations, international organizations, civil affairs units, special forces). Working as a social scientist on a Human Terrain Team offers a rare and unique opportunity to help reshape the military's execution of their mission by offering them a much greater appreciation of existing socio-cultural realities and sensitivities in the countries where they are operating. This position also offers an opportunity to develop new methods for data collection and analysis. Social scientists will be able to write about their experiences and otherwise contribute to the academic literature in their field after participation, subject to standard security review.

The characterization of the work offered by the hiring entity seems meticulously crafted to meet the requirements of AAA Code of Ethics restrictions, including even language re-introduced in 2009 to promote "transparency" and inhibit "secret research." The ad deftly addresses suspicions or reservations about the nature of the work the recruited candidate might finally be ask to undertake, or the restrictions under which he or she might finally be forced to work. As we have seen, however, living up to the requirements or restrictions of that AAA Code will likely prove to be the least of a job applicant's legitimate worries.

Job applicants in this instance are required to hold US citizenship, and to hold or have the ability to obtain and maintain a security clearance. The jobs also require a PhD in anthropology, or in a "related field such as sociology, political science, history, theology [sic], economics, public policy, social psychology, or area studies." Despite reports of salaries in excess of $20,000 per month during deployment, these positions are not for just anyone. In fact, one of the main criticisms of the program is that (apart from the fictional Hollywood character, "Indiana Jones") few if any persons are likely to possess the entire skill set required. As if this weren't enough, the BAE Systems advertisements also warn that successful candidates must be able to endure rigorous conditions and physical hardships of the sort that would lead even the "early," let alone the "later" Indiana Jones to complain, including:

- Adverse battlefield conditions.
- Heat well in excess of 110 degrees in the Summer, cold and freezing conditions during the winter.
- Rough Terrain including climbing rocks, mountains and fording bodies of water.
- Hostile environment to include persons that may cause bodily harm, injury or loss of life.
- Required to work at times with little sleep or rest for extended periods of time while producing both physically and mentally challenging projects.
- Extended travel by foot, military ground vehicles and air transport into mountainous regions for extended periods of time.
- Sleeping on the ground in environmentally unprotected areas.
- Required to lift 40–75 pounds of equipment and personal gear including protective equipment several time a day.

- Required to carry 40–75 pounds of gear, personal protective equipment for 10–16 hours a day while walking in rough terrain.

Finally, the ads suggest that preferred candidates will, not surprisingly have experience living or working in the Middle East and possess language skills in Arabic, Pashtun or Dari.

Assuming that BAE contract staff (or, more recently, Army TRADOC staff) are fortunate enough to locate and recruit interested individuals with these impressive skill sets, what positions might they occupy? Typical positions include HTS Team Leader, analyst, research manager, and a more general opening for a "social scientist." Here is where the nature of the team structure and organization come into play. The "Human Terrain Team Leader" requires *prior substantial military experience* as a brigade staff officer, at the rank of O-5 (i.e., Lieutenant Colonel or Navy Commander) or above, and preferably, graduation from a Staff and Command College (or its equivalent). The HTS Team Leader (the BAE job advertisement states), "will lead the Human Terrain System (HTS), which is a team that will collect and analyze data from the Brigade Combat Team (BCT) and Regional Combat Team (RCT) to obtain cultural and political awareness in order to sustain and foster stabilization. The HTS project is designed to improve the *gathering, interpretation, understanding, operational application and sharing of local population knowledge* [my emphasis] at the BCT and RCT and Division levels. The Human Terrain Team Leader will be the BCT commander's principal human terrain advisor, responsible for supervising the team's effort and *helping integrate data into the staff decision process*. The key attribute of the HTT team leader is the ability to successfully integrate the HTT into the process of the BCT in an effective and credible manner and become a trusted advisor to the BCT commander."

An HTS Analyst is also required to have prior military experience, preferably with intelligence debriefing. The analyst "will be a part of the Human Terrain System (HTS), which is a team that will collect and analyze data from the Brigade Combat Team (BCT) and Regional Combat Team (RCT), once again, to obtain cultural and political awareness in order to sustain and foster stabilization. The Human Terrain Analyst will serve as the primary human terrain data researcher. He/She will participate in debriefings, and will interact with other target area organizations and agencies." Yet another position, the "Human Terrain Research Manager" must also have some prior military experience, preferably in HUMINT (human intelligence), and is expected to integrate the team's findings into a broader, regional context "to obtain cultural and political awareness in order to sustain and foster stabilization."

The position for a "Social Scientist" as a member of the HTS team is the *only position that does not require prior military service*, and which emphasizes in particular the skill set acquired in earning a doctorate in anthropology.

The Social Scientist will provide local interpretation of socio-cultural data, information and understanding of local and regional culture. The Social Scientist will work closely with and possess similar skills as the Cultural Analyst, but with more focus on the larger region in which the target area is embedded.

Note that this last was the position that both Michael Bhatia and Paula Loyd apparently held in their respective teams in Afghanistan. Lt. Col. Edward Villacres, with the 82 Airborne Division HTS team in Iraq, was the "HTT team leader," while Nicole Suveges, also in Iraq, was an HTS analyst. Both these individuals had prior military experience at the officer and enlisted level, respectively.

There are a number of grave problems with these job descriptions, most of which are not the "questions of professional ethics" envisioned by the HTS program's critics back in the US. But those problems are formidable nonetheless, and may raise a host of moral and ethical questions in their own right. Note, for example, that all members of the proposed HTS team except the subject matter expert in social science must have substantial prior military experience. Thus, the abilities and training in surviving wartime conditions that might be envisioned, and even the ability to work cooperatively with brigade and regional combat forces, are likely to be handled by those with prior experience in such operations, with the academic anthropologist "tagging along." Neither Paula Loyd, nor Nicole Suveges, nor Michael Bhatia, for example, were neophyte anthropology scholars lured in by high salaries, or minimally trained at Fort Leavenworth, and then released recklessly into a combat zone, as was erroneous reported at first. Instead (although this does not mitigate the tragedy of their loss), two were prior military enlisted personnel with considerable experience in the country (Suveges, Loyd), while Bhatia was a civilian with substantial experience in counterinsurgency and post-conflict interventions. They are the ones whose job it is to liaise with brigade and regional commanders, coordinate embedded team activities with combat and security operations, and, in all likelihood, keep their less-well-oriented academic "SME" (subject matter expert, the "adjunct professor of anthropology" from Cal State cited above, for example) from getting his or her head blown off (or, just as seriously, causing others to get their heads blown off by being in the wrong place at the wrong time).

This is not an ideal working relationship. Military commanders in the field with some experience of these kinds of arrangements opine that the likely problems with HTS teams is not that they will provide "critical or clandestine intelligence" about local populations (which might better be obtained from the brigade's assigned native translator), let alone that they will be asked to collude in doing harm to locals. Rather, having all these "non-combat personnel" wandering around in a combat zone is a dangerous distraction for all concerned. *Pravda* journalist, John Stanton's hearsay report of opinions that "someone is likely to get killed" over HTS[20] pertains more

---

20   An earlier article dated July 23, 2008 at the same hyperlink is the first of three published by this journalist, most simply repeating, updating, or amplifying the charges

to that problem than to the alleged lack of training of the prior military personnel who form the bulk of the team.[21]

The risk might be worthwhile, if the subject expertise that the academic social scientist could bring to bear on brigade operations were substantial and accurate. But, as the BAE advertisement suggests, there is very little likelihood of that. Given that all these regions have been in a state of armed conflict for over two decades, it is extremely unlikely that any scholar who was not native-born (and perhaps a refugee from the violence) would have found the opportunity of acquiring remotely the diverse kinds of skill sets requested. Bhatia and Loyd were notable and commendable exceptions in this respect. As the BAE advertisement also makes clear, the hiring contractor has adopted (shall we say) a rather ecumenical, eclectic attitude toward relevant subject matter expertise: if the candidate lacks a degree in anthropology, a degree in theology, economics, or (only slightly more plausibly) social psychology will suffice.

Where's the Outrage?

Once again, most of the AAA debate has focused on the principled outrage within the discipline of anthropology occasioned by such ads, and by the revelations that programs like HTS and companies like BAE Systems are recruiting academically-trained anthropologists for such work. Is the concern warranted? Is there, finally, any good reason, in principle, why an anthropologist should not accept these jobs?

On one hand, I am inclined at this point to conclude that the answer to that question, grounded in principle, and based upon the foregoing job descriptions, is simply "no." On the other hand, from the foregoing accounts, this does not seem to be quite the right question, not at least for HTS team members engaged in HTS$_1$. The proper question is, could anyone hired through such a process meet these requirements successfully? Would their brief, four-month orientation at Fort Leavenworth be enough (absent a prior career either in the military itself, or as a field anthropologist under the most demanding of circumstances) to prepare them for the risks and rigors they would face? Would they end up proving an asset, or a liability, even to their HTS team members, let alone to the military forces they are trying to assist? Would the rewards they might bring in protecting human subjects

---

found here. Stanton describes himself as "a Virginia-based writer specializing in national security and political matters".

21   A recent report by *Washington Post* correspondent Vanessa Gezari (Gezari, 2009, p. 20), describes the work of psychologist and conflict resolution specialist, Dr. Karl Slaikeu, Paula Loyd's replacement on her former Human Terrain Team. Reconnoitering in the village of Pir Zadeh in southern Afghanistan, the Brigade Combat Team commander, 24-year-old Army Lt. Terrence Paul Dunn, is obliged to issue instructions to this inexperienced HTS team member on how to stay out of danger (and out of the way) if the unit should come under fire.

and lowering the incidents of conflict or mistaken resort to deadly force be worth the risks they would incur to themselves, let alone worth the implicit threat that all non-combatant support personnel pose to military units when deployed in a combat zone? That is to say, the real problems, including true moral dilemmas or conflicts of professional ethics, likely arise in filling these positions, providing the promised training, and delivering on the promises contained in the description of the overall HTS mission.

At present, we are reliant for our evaluation of such important questions entirely on the scattered reports and rumors emanating from the field. As noted, the scant factual reports vary widely. In this respect, the testimony of our embedded anthropologist, Mark Dawson, "blogging" in Iraq, speaks forcefully, and seems sharply at odds with what both supporters and critics of the programs have attempted to project as its current condition. He writes:

> What's it like? The Army has been helpful, hospitable, arranged about any kind of fieldwork we wish to do. It should be no surprise that they were polite and helpful, but deeply skeptical over any real value we might add other that eating the food in the Dining Facility (DFAC). But now we have gotten a seat at the table so to speak, people request our assistance on various topics.
>
> Looking at the AAA controversy after two months actually in the field makes some of the issues laughable. This idea that Human Terrain Teams are involved in gathering intelligence for example (we'll ignore for the moment the error of jargon people make: you don't 'gather' intelligence). I have seen this written about as if there are people in trench coats and dark glasses hovering by the HTT door just dying for a chance to peek at our work! Lord what a load of nonsense. I don't know how many times this can be said, no … we don't, ever. They don't even want us to. Why, because it's not our jobs, and they have professionals for that.
>
> [Instead] we look at people in transition from jobs in one sector to another, how effective governance is in areas, issues of economy. The meetings I sit in are about building schools, putting in water purification facilities, trying to help local governments get more support for their communities from the Iraq national government, understanding complex issues related to agriculture and the economy … any anthropologist involved with development work would be pretty [much] at home here.

Once again, it is important to stress that this account, while it may count as "field notes" of a sort, is the report of a single individual. It may not be representative of successes or failures of the program as a whole, let alone of the adequacy with which this kind of program is administered by the private sector in the midst of combat. What is telling, and what rings true on the basis of the testimony of myriad others with "on-site, on-the-ground" experiences of this sort to report from their deployment in both Afghanistan and Iraq (including skeptical journalists embedded with combat troops) is what follows:

I think what has stuck me the most since I have been here are the relationships that the military have built with their Iraqi counterparts helping them [in] creating their own local governmental structures. Most of these young women and men have been trained to lead tanks and infantry companies. They showed up in Iraq and someone told them to establish security in an area and help the local population create an effective structure for representation to the provincial and national level. I have done part of my fieldwork in these meetings all over our area of the country. It's amazing and impressive. They have made it happen with no training and little support. Its fragile, rough, does not work always the way it should, but it's indeed on the way. These military people figured out how to make it happen simply because they had no other choice but to try. *They have all told me they wish anthropologists and governmental experts had been there to help them 12 months ago.* [my emphasis]

This recent account squares, in its mixed assessment and moral ambiguity, with similar accounts of progress or its lack from military personnel rotating in from the field. Surface Warfare officers, with degrees in oceanography, find themselves detailed to disassemble IEDs in the "Sunni Triangle," while P-3 pilots and submariners "I.A'd" ("individually augmented," in military-speak) to Afghanistan, command special forces units in rebuilding schools, roads, hospitals, and constructing wells and water treatment plants while fending off Taliban attacks against the villagers in their regions. "It's okay, sir," one Navy Captain joked recently, during a video teleconference from Kabul with the US President, Secretary of State, and Secretary of Defense. "I spent the night in a Holiday Inn Express."[22]

Dawson's description also aligns with the earlier accounts of independent news reports, like those of David Rodhe of the *New York Times*, and also by the head of the HTS team attached to the 82nd Airborne division in Iraq in the Fall of 2007 on the "Diane Rehm Show." In all instances, those reports are more detailed, nuanced, and usually at odds with the portraits painted by war's supporters, detractors, and also with the verbal "snapshots" obtained from conventional media sources.[23] The complexity of which this individual testifies, and the bureaucratic and logistical complaints he lodges, can be traced to a number of factors, worthy of consideration because they bears substantively on whether, on balance, the HTS program will prove effective, let alone morally justifiable, in achieving its aims.

---

22   That story was related during a brief furlough in the US by Captain Scott Cooledge, USN, commander of a provincial reconstruction team in Gahzni Province, Afghanistan, the keynote speaker at the Stockdale Center's semi-annual Ethics Essay Award banquet on September 17, 2008. Secretaries Gates and Rice, he reported, were mystified by his reference to a US hotel chain in Afghanistan, but the President got the joke.

23   For more nuanced background accounts of Iraq, in particular, readers might wish to consult the reflections published in the Thanksgiving Day, 2008 *Washington Post*, by Captain Giles Clarke, USMC, midway through his third tour of duty since 2003, with a personal investment of over 19 months in that country (Clarke, 2008).

What emerges as the principle moral (as opposed to ethical) dilemma on the foregoing accounts has less to do with "clandestine research" or malevolent covert actions, than with safety and transparency. The HTS program is similar, in many respects, to the practice of embedding journalists, chaplains, and perhaps also medical personnel with combat troops. Such personnel can be tremendously helpful, but they also pose a grave danger to themselves and others. The kidnapping of reporter David Rohde, and the serious injury and subsequent death of HTS team member Paula Loyd, illustrate the problem of "forward-deploying" non-essential personnel in combat zones. The moral dilemma, as stated, is whether the benefits and rewards of doing so outweigh these risks.

It is impossible to answer that question, in turn, without a much fuller account of HTS activities that we are thus far able to give. These important questions of safety and cost-effectiveness can only be addressed through a full program review. The Human Terrain System is, admittedly a fairly new program, with only a few units in the field, and little assessment thus far of its overall effectiveness. That situation, in turn, owes much to the unusual requirements of the program, recruitment difficulties, and complexity of operation. This would be true even if the program was able to retain the very best talent, and train and deploy individuals with the most exemplary personal character on all fronts. The problems would be magnified, quite naturally, if some of the personnel hired and deployed turned out to be unqualified, irresponsible, or incompetent, or if (as alleged) some management staff themselves were irascible, difficult, or malevolent.

In its magnitude, this requirement is a far greater problem in practical or applied ethics than any malevolent actions or intentions, or any "secret or clandestine research" in which participants may be engaged. As is typical in irregular warfare generally, we have, through this experimental program, apparently recruited some very decent and talented individuals, placed them in harm's way in the midst of combat, and asked them to "do some good." Like our military forces, to whom we gave essentially the same vague IW mission, we have provided no guidance, no framework for success, and no procedures to evaluate how well or poorly they are doing. Meanwhile, back home, uninformed pundits and journalists spin dark tales of malevolent conspiracy, and rant ponderously and pompously about "secret research." Mark Dawson is right: it would all be laughable, were it not—as in the cases of Bhatia, Suveges, and Paula Loyd—so thoroughly tragic.[24]

---

24 Psychologist and conflict resolution specialist, Dr. Karl Slaikeu, Paula Loyd's HTS replacement in southern Afghanistan, recently confirmed this analysis. He describes his own misgivings and awareness of the ethics debate in the AAA over the propriety of HTS. During his training at Ft. Leavenworth, and subsequently during his deployment in Maywand, Southern Afghanistan, he describes himself as watching "for anything that might jeopardize ethical standards by endangering local people." "It just hasn't come, he said, "and I've been looking for it" (Gezari 2009, 23).

References

AAA (2007), "American Anthropological Association Executive Board Statement on the Human Terrain System Project," 31 October. Available online at: http://dev.aaanet.org/issues/policy-advocacy/Statement-on-HTS. cfm (Last accessed: 27 November, 2008).

Avant, D. (2002), "Privatizing Military Training: A Challenge to US Army Professionalism," in Snider, D.M. and Watkins, G.L. (eds.), *The Future of the Army Profession* (New York: McGraw-Hill), 179–96.

Bender, B. (2007), "Efforts to Aid US Roil Anthropology," *The Boston Globe*, 8 October 2007.

Berreman, G. (1971), "Ethics, Responsibility, and the Funding of Asian Research," *Journal of Asian Studies*, 30:2.

Berreman, G. (2003), "Ethics Versus Realism in Anthropology: Redux," in Fluehr-Lobban, C. (ed.), *Ethics and the Profession of Anthropology*, 2nd edn (Walnut Creek, CA: AltaMira Press), 51–83.

Caplan, P. (ed.) (2003), *The Ethics of Anthropology: Debates and Dilemmas* (London: Routledge).

Carafano, J. (2008), *Private Sectors, Public Wars: Contractors in Combat* (Westport, CT: Praeger Publishers).

CEAUSSIC (2007), "Final Report of the AAA Ad Hoc Commission on the Engagement of Anthropology with US Security and Intelligence Communities," 4 November 2007 Available online at: http:www.aaanet.org/pdf/Final_Report.pdf (Last accessed: 27 November 2008).

Clarke, G. (2008), "Going the Distance in Iraq," *The Washington Post* (27 November): A29.

Condominas, G. (1994), *We Have Eaten the Forest: The Story of a Montagnard Village in the Central Highlands of Vietnam* (New York: Kodansha).

Constable, P. (2009), "A Terrain's Tragic Shift: Researcher's Death Intensifies Scrutiny of US Cultural Program in Afghanistan," *The Washington Post* (18 February): C1, C8.

Fainaru, S. (2008a), *Big Boy Rules: America's Mercenaries Fighting in Iraq* (Philadelphia, PA: Da Capo Press).

Fainaru, S. (2008b), "Soldier of Misfortune," *The Washington Post* (1 December): C1–C2.

Ferguson, R. and Whitehead, N. (1992) *War in the Tribal Zone: Expanding States and Indigenous Warfare* (Santa Fe, NM: School of American Research Press).

Fluehr-Lobban, C. (ed.) (1994), "Informed Consent in Anthropological Research: We Are Not Exempt," *Human Organization*, 52:1: 1–10. Reprinted in Fluehr-Lobban, C. (2003) 159–77.

Fluehr-Lobban, C. (2003), *Ethics and the Profession of Anthropology*, 2nd Edn, (Walnut Creek, CA: AltaMira Press).

Fluehr-Lobban, C. (2003a), "Ethics and Anthropology: 1890–2000," in *Ethics and the Profession of Anthropology*, 2nd Edn (Walnut Creek, CA: AltaMira Press). 1–28.

Fluehr-Lobban, C. (2003b), "Darkness in El Dorado: Research Ethics, Then and Now," in *Ethics and the Profession of Anthropology*, 2nd Edn (Walnut Creek, CA: AltaMira Press), 85–106.

Fluehr-Lobban, C. (2003c), "Informed Consent in Anthropological Research: We are Not Exempt," in *Ethics and the Profession of Anthropology*, 2nd Edn (Walnut Creek, CA: AltaMira Press) 159–77.

Fluehr-Lobban, C. (2008), "New Ethical Challenges for Anthropologists: The terms 'Secret Research' and 'Do No Harm' need to be clarified," *The Chronicle of Higher Education Review*, 55:12, November: B11.

Frese, P.R and Harrell, M.C. (eds) (2003), *Anthropology and the United States Military: Coming of Age in the Twenty-First Century* (New York: Palgrave Macmillan).

Fosher, K. (2010), "Yes, Both, Absolutely: A Personal and Professional Commentary on Anthropological Engagement with Military and Intelligence Organizations," Invited chapter in the *Anthropology and Global Counterinsurgency* (Chicago: University of Chicago Press).

Fosher, K. (2009), *Under Construction: Making Homeland Security at the Local Level* (Chicago: University of Chicago Press).

Fujimura, C. (2003), "Integrating diversity and understanding the Other at the U.S. Naval Academy," in Rese, P. and Harrell, M.C. (eds), *Anthropology and the United States Military: Coming of Age in the Twenty-first Century* (New York: Palgrave Macmillan), 147–51

Gezari, V. (2009), "Rough Terrain," *The Washington Post Magazine* (30 August), 16–29.

Glenn, D. (2009), "Anthropologists Adopt New Language Against Secret Research," *Chronicle of Higher Education* (19 February 2009).

Gonzalez, R.J. (2007), "We Must Fight the Militarization of Anthropology," *Chronicle of Higher Education Review*, 53:22, 2 February 2007, B20.

Greenburg, K. (2009), "When Gitmo Was (Relatively) Good," *The Washington Post Outlook* (25 January 2009): B1, B4.

Gusterson, H. (2007), "Anthropology and Militarism," *Annual Review of Anthropology*, 36, 155–75.

Hersh, S.M. (2004), "The Grey Zone," *The New Yorker* (24 May).

Horowitz, I.L. (1967), *The Rise and Fall of Project Camelot: Studies in the Relationship between Social Science and Practical Politics* (Cambridge, MA: MIT Press).

Jaschik, S. (2009), "Anthropologists Toughen Ethics Code," *Inside Higher Ed.* (19 February). Avhttp://www.insidehighered.com/news/2009/02/19/anthro.

Kilcullen, D. (2009), *The Accidental Guerilla: Fighting Small Wars in the Midst of a Big One* (Oxford: Oxford University Press).

Lesser, F. (1981), "Franz Boas," in Sydel Silberman (ed.), *Totems and Teachers: Key Figures in the History of Anthropology*, 2nd edn (Walnut Creek, CA: AltaMira Press, 2004) 1–25.

Lucas, G.R. (2008), "The Morality of Military Anthropology," *Journal of Military Ethics*, 7:3: 165–85.

Lucas, G.R. (2009), *Anthropologists in Arms: the Ethics of Military Anthropology* (Lanham, MD: AltaMira Press).

McFate, M. (2005a), "Anthropology and Counterinsurgency: The Strange Story of their Curious Relationship," *Military Review* (March–April).

McFate, M. (2005b), "The Military Utility of Understanding Adversary Culture," *Joint Forces Quarterly*, 38 (3rd Quarter): 42–7.

Mead, M. (2000 [1942]), *And Keep your Powder Dry: An Anthropologist Looks at America* (New York: Berghahn Books).

Mintz, R. (1981), "Ruth Benedict," in Sydel Silberman (ed.), *Totems and Teachers: Key Figures in the History of Anthropology*, 2nd edn. 2004 (New York: AltaMira Press) 103–24.

Montreaux (2008), "Montreux Document on Pertinent International Legal Obligations and Good Practices for States Related to Operations of Private Military and Security Companies during Armed Conflict," (Montreux, Switzerland: International Committee of the Red Cross) (17 September 2008) Available online at: http:www.icrc.org/web/eng/siteeng0.nsf/htmlall/montreux-document-170908/$FILE/Montreux-Document.pdf (Last accessed: 27 November 2008).

Nature (2008), "Failure in the Field: The US Military's Human Terrain Programme Needs to be Brought to a Swift Close," *Nature*, 456 (11 December), 676.

NCA (2007), "Pledge of Non-Participation in Counterinsurgency," *Network of Concerned Anthropologists*. Available online at: http:concerned. anthropologists.googlepages.com (Last accessed: November 27, 2008).

Nuti, P.J. (2006), "Smart Card: Don't Leave Military Base Without It," *Anthropology News*, October 6, 2006: 15–16.

Nuti, P.J. and Fosher, K. (2007), "Reflecting Back on a Year of Debate With the Ad Hoc Commission," *Anthropology News*, 48:7: 3–4.

Packer, G. (2006), "Knowing the Enemy," *The New Yorker*, 16 December: 60–9.

Pels, P. (1999), "Professions of Duplexity," *Current Anthropology* (April): 40:2.

Pels, P. and Salemink, O. (2000), *Colonial Subjects: Essays on the Practical History of Anthropology* (Ann Arbor, MI: University of Michigan Press).

Price, D. (1997), "Anthropological Research and the Freedom of Information Act," *Cultural Anthropology Methods*, 9:1: 12–15.

Price, D. (1998a), "Gregory Bateson and the OSS: World War II and Bateson's Assessment of Applied Anthropology," *Human Organization*, 75:4: 379–84.

Price, D. (1998b), "Cold War Anthropology: Collaborators and Victims of the National Security State," *Identities*, 4 :3–4: 389–430.

Price, D. (2003), "Anthropology sub rosa: The CIA, the AAA, and the Ethical Problems Inherent in Secret Research," in Fluehr-Lobhan, C. (ed.), *Ethics and the Profession of Anthropology*, 2nd Edn (Walnut Creek, CA: AltaMira Press) 29–49.

Price, D. (2004), *Threatening Anthropology: McCarthyism and the FBI's Surveillance of Activist Anthropologists* (Durham, NC: Duke University Press).

Price, D. (2008), *Anthropological Intelligence: The Development and Neglect of American Anthropology in the Second World War* (Durham, NC: Duke University Press).

Price, D. and Gonzalez, R.J. (2007), "When Anthropologists Become Counterinsurgents," in Alex Cockburn and Jeffrey St. Clair (eds), *Counterpunch*, 27 September. Available online at: http:www.counterpunch. org/gonzalez09272007.html (Last accessed: 27 November 2008).

Petraeus, D. and Amos, J.F (eds) (2007), "Counterinsurgency," *Army Field Manual* (Washington, DC: US Government Printing Office) 3–24.

Rehm, D. (2007), "Anthropologists and War," Guest host, Susan Page, *USA Today*. The Diane Rehm Show. WAMU Public Radio, Washington DC. Wednesday, 10 October 2007 at 10–11 a.m. EDT Available online at: http:wamu.org/ programs/dr/07/10/10.php (Last accessed: 27 November, 2008).

Rohde, D. (2007), "Army Enlists Anthropology in War Zones," *New York Times*, 5 October 2007. Available online at: http://www.nytimes.com/2007/10/05/ world/asia/05afghan.html?pagewanted=1&_r=1&sq=David%20Rohde&st=cs e&scp=2 (Last accessed: 27 November 2008).

Rubinstein, R.A. (1988), "Anthropology and International Security," in R.A Rubinstein and M.L Foster (eds), *The Social Dynamics of Peace and Conflict* (1997) (Dubuque, IA: Kendall/Hunt Publishing Co.), 17–34.

Rubinstein, R.A. (2001), *Doing Fieldwork: the Correspondence of Robert Redfield and Sol Tax* (New Brunswick, NJ: Transaction Books).

Rubinstein, R.A. (2003), "Politics and Peace-keepers: Experience and Political Representation among United States Military Officers," in P.R Frese and M.C Harrell (eds), *Anthropology and the United States Military: Coming of Age in the Twenty-first Century* (New York: Palgrave Macmillan), 15–27.

Rubinstein, R.A. (2006), "Anthropology, Peace, Conflict, and International Security," San Jose, CA, 106th Annual Meeting of the American Anthropological Association.

Rubinstein, R.A. (2008), *Peacekeeping Under Fire: Culture and Intervention* (Boulder, CO: Paradigm Publishers).

Rubinstein, R.A. (2008a), "Ethics, Engagement, and Experience: Anthropological Excursions in Culture and the Military," Seminar on "Scholars, Security and Citizenship," 24–25 July 2008 (Santa Fe, NM: School of Advanced Research on the Human Experience).

Sahlins, M.D. (1965), "The Established Order: Do Not Fold, Spindle, or Mutilate," A speech delivered at the November 1965 annual meeting of the American Anthropological Association. Reprinted in Irving Louis Horowitz (ed.), *The*

Rise and Fall of Project Camelot: Studies in the Relationship between Social Science and Practical Politics (Cambridge, MA: MIT Press), 71–9.

Salmoni, B.A. and Holmes-Eber, P. (2008), *Operational Culture for the Warfighter: Principles and Applications* (Quantico, VA: Marine Corps University Press).

Selmeski, B.R. (2007a), "Who Are the Security Anthropologists?," *Anthropology News* 48:5 (May 2007): 11–12.

Selmeski, B.R. (2007b), "Military Cross-Cultural Competence: core concepts and individual development," Centre for Security, Armed Forces & Society, *Occasional Paper Series*, 1 (Kingston, Ontario: Royal Military College of Canada, 16 May).

Selmeski, B.R. (2008), "Anthropology for the (Military) Masses: A Moral-Practical Argument for Educational Engagement," San Francisco, CA: 107th Annual Meeting of the American Anthropological Association (22 November).

Schachtman, N. (2008), "Montgomery McFate: Use Anthropology in Military Planning," *Wired Magazine*, 16:10 (September 22). Available online at: http://www.wired.com/politics/law/magazine/16-10/sl_mcfate (Last accessed: 27 November 2008).

Schachtman, N. (2008a), "Army Social Scientist Set Afire in Afghanistan," *Wired Magazine/Danger Room* (6 November 2008). Available online at: http://blog.wired.com/defense/2008/11/army-social-sci.html (Last accessed: 27 November 2008).

Schachtman, N. (2009), "'Human Terrain' Contractor Guilty of Manslaughter," *Wired Magazine/Danger Room* (3 February 2009). Available online at: http://blog.wired.com/defense/2009/02/human-terrain-c.html. Accessed 9 February 2009.

Schachtman, N. (2009a), "Human Terrain Contractors' Pay Suddenly Slashed," *Wired Magazine/Danger Room* (13 February 2009). Available online at: http://blog.wired.com/defense/2009/02/more-hts-mania.html.

Silverman, S. (ed.) (2004), *Totems and Teachers: Key Figures in the History of Anthropology*, 2nd edition (New York: AltaMira Press).

Silverman, S. and Métraux, R. (1981), "Margaret Mead" in Sydel Siverman (ed.), *Totems and Teachers: Key Figures in the History of Anthropology*, 2nd edn (New York: AltaMira Press, 2004), 198–221.

Silvert, K.H. (1965), "American Academic Ethics and Social Research Abroad: the Lesson of Project Camelot," *American Universities Field Staff Reports (West Coast South American Series)*, 12:3 (July 1965). Reprinted in Irving Louis Horowitz (ed.), *The Rise and Fall of Project Camelot: Studies in the Relationship between Social Science and Practical Politics* (Cambridge, MA: MIT Press), 80–106.

Simons, A.J. (1997), *The Company They Keep: Life Inside the US Army Special Forces* (New York: The Free Press).

Singer, P.W. (2008), *Corporate Warriors: The Rise of the Privatized Military Industry*, updated edition (Ithaca, NY: Cornell University Press).

Stanton, J. (2008), "US Army Human Terrain System in Disarray: Millions of Dollars Wasted, Two Lives Sacrificed," *Pravda*, July 23, 2008. English language version available online at: http://english.pravda.ru/topic/Human_Terrain_System-607 (Last accessed: 27 November 2008).

Stanton, J. (2009), "US Government Takeover of Human Terrain System," *Pravda* (11 February). Available online at: http://cryptome.info/0001/hts-bailout.htm

Stockman, F. (2009), "Anthropologist's War Death Reverberates," *The Boston Globe* (12 February 2009). Available online at: http:www.boston.com/news/world/middleeast/articles/2009/02/12/anthropologists_war_death_reverberates/?page=1 (Last accessed: 9 March 2009).

Sullivan, J.P. (2009), "Partnering with Social Scientists," *Marine Corps Gazette*, 93:1: 53–7.

Tiron, R. (2009), "Panel Says Pentagon Relies Too Heavily on Contractors," *The Hill* (12 February). Available online at: http://thehill.com/the-executive/panel-says-pentagon-relies-too-heavily-on-contractors-2009-02-12.html (Accessed 21 February 2009).

Vergano, D. and Weise, E. (2008), "Should Anthropologists Work Alongside Soldiers?," *USA Today* (9 December 2008): 5D.

Wakin, E. (2007 [1992]), *Anthropology Goes to War: Professional Ethics and Counterinsurgency in Thailand* Monograph #7, University of Wisconsin Center for Southeast Asian Studies (Madison, WI: University of Wisconsin Press).

Winnick, D. (2008), "Anthropology and the Military: A Summary of Related 2007 Annual Meeting Events," *Anthropology News*, no. 1. January 2008: 18–19.

Wolf, J.G and Jorgensen, E.R. (1970) "Anthropology on the Warpath in Thailand," *New York Review of Books*, 15: 26–35.

Chapter 10

To Whom does a Private Military Commander Owe Allegiance?[1]

Deane-Peter Baker

A concern regularly expressed by critics of the trend towards the outsourcing of traditional military functions to private contractors is that these contractors, or perhaps more importantly the managers ("commanders") among them, ultimately owe allegiance to their company's owners or stock holders. Private contractors ought, therefore (so the argument goes), not be trusted. In this chapter I explore this claim through the lens of recent civil–military relations theory, and conclude that, with one important exception, there is no reason to think that private contractors are in general fundamentally any less reliable than their uniformed counterparts. The exception to my thesis applies to those cases where military personnel might reasonably be expected to sacrifice their lives in the service of their nation. Under those conditions, I argue, it would be unreasonable (and unwise) to rely upon a contractor to serve in the same way as a soldier, sailor, marine or airman. The flip side of this, I argue, is that there are missions in which there is a moral duty to prefer the employment of contractors to uniformed personnel. This applies particularly to those discretionary operations that serve no clear national interest but which are undertaken in the service of humanity.

Introduction

Although the private provision of military services is by no means a new phenomenon, what is new in our era is the fact that many Western governments are increasingly seeing private military companies as legitimate service providers. This shift has troubled many commentators in academia and the policy world, for a range of different reasons. One of the central concerns that is regularly voiced is the worry that private companies that provide military services owe their primary allegiance to the owners or stock holders of those companies, and

1 A much shorter version of this chapter was first presented at the 2nd Annual Defence Ethics Conference held at the Defence Academy, Shrivenham, UK in December 2008. I am grateful for comments received on that occasion, to David Weatham for inviting me and organizing a great event, and to the British Academy for the financial support that enabled my attendance at the conference.

that this makes them untrustworthy. At the strategic level, the concern is that the outsourcing of traditional military functions into private hands could potentially undermine civil–military relations. Given the traditional view of civil–military relations, one defined by Samuel Huntington in his classic book *The Soldier and the State*, it is hardly surprising that this should be a cause for concern. Following this orthodoxy, the central determinants of appropriate civil–military relations are "soft" factors, primarily such features as military professionalism, honor and a culture of submission to civil authority. Clearly, while private military companies may sometimes display these features, there is no particular reason to think that they will. Elke Krahmann expresses this concern well when she writes that,

> High levels of mutual interpenetration and common identity are regarded as supportive of democratic civil–military relations because they increase the commitment of the armed forces to the defence of their society, ensure that the military shares societal norms and beliefs, and facilitate support of government policies. The model of the private military contractor undermines these historically established means of democratic control in a number of ways and thus requires a revision of traditional civil–military relations. (Krahmann 2008, 258)

Huntington's approach has, in recent times, been challenged by a new theory of civil–military relations, one advocated by Peter Feaver. Feaver's "Agency Theory," while recognizing the importance of the soft factors so central to Huntington's approach, also recognizes the importance a range of "hard" determinants of whether military forces submit to democratically elected civilian leadership. It is one of the main contentions of this chapter that Agency Theory provides a superior analytic tool for addressing the question of appropriate relations between private military companies and democratically elected governments. I will attempt to show that, rather than the blanket condemnation of private military companies that seems to result from the application of Huntington's model of civil–military relations, Agency Theory provides an agile tool that can advance the debate over military privatization by giving a significantly more fine-grained account of what is necessary for appropriate civil–private military relations. As Peter W. Singer points out, this kind of analysis is largely missing from the academic literature:

> From its very beginning, the underlying basis of current civil–military relations theory has been fairly simple. Essentially it is a story of balancing proper civilian control with the military professionals' need for autonomy to do their jobs properly. Although ongoing debates over where exactly these lines of control should be drawn, the whole of civil–military relations theory, regardless of its viewpoint, sticks to this general assumption of a dualistic balance between soldiers and state. Presently, civil–military relations theory does not fully account for any potential role of external, third-party influences on this two-sided structure. (Singer 2003, 196)

It is critical that this analysis be carried out, in order to ensure that we avoid falling into the trap, pointed out by Musah and Fayemi, of those scholars "who see the use of today's mercenaries (sic) as the effective antidote for insecurity in zones of complex emergencies, but pay little or no attention to the subversion of the very state sovereignty the mercenaries claim to protect" (Musah and Fayemi 2000, 27).

Before continuing any further, it is necessary to make a few comments on the parameters of this chapter. First, it must be acknowledged that the vast majority of the contractors that are present on today's battlefields are unarmed and serve primarily in support (especially logistics) roles. Those that do carry and occasionally use firearms dislike the label private *military* contractors, preferring instead the term private security contractors, or something similar. In this chapter I am not intending to specifically address any existing group of contractors. Instead I am deliberately focusing on the idea of contractors serving what we might refer to as "fully fledged" military functions. In recent history it is arguably only the now defunct firm Executive Outcomes that have come close to what I have in mind. The purpose of this chapter is not, I repeat, to address any particular group of contractors operating today, but instead to lay a theoretical framework within which to conceptually understand the idea of *private military* obedience to elected civilian principals.

Feaver's Agency Theory

I have sought, elsewhere, to make the case for preferring Peter Feaver's Agency Theory over other theoretical approaches to civil–military relations (Baker 2007), and space does not allow for a rehearsal of that argument here. I begin, instead, with a brief summary of Feaver's theory, and then move on to assess the nature of civil– (private) military relations in its light.

Feaver's model draws on principal agent theory, a framework widely used in economic and political analysis. Its goal is to address problems of agency, particularly between actors in a position of superiority or authority (principals) and their subordinates (agents). The classic case is perhaps the employer–employee relationship. In such cases the goal of principal–agent theory is to address the problem of how the employer ensures that the employee does what is required of her, or in other terms, how the employer ensures that the employee is "working" rather than "shirking." Feaver argues that civil–military relations can be seen as "an interesting special case" (Feaver 2003, 12) of the principal–agent relationship. Because this "special case" has features that are unique and not broadly applicable to other principal–agent relationships, Feaver coins the term "Agency Theory" to describe it (Feaver 2003, 55).

At the heart of Agency Theory is the idea that civil–military relations is essentially a form of strategic interaction between civilian masters (principals) and their military servants (agents). In that strategic interaction civilians choose methods by which to monitor the military. What methods are chosen depends on what expectations the civilians have about the degree to which the military will submit to their authority. Submission or obedience, is, in Feaver's terminology, "'working', while rebellion or refusal to obey is 'shirking.'" "The military decides whether to obey in this way, based on military expectations of whether shirking will be detected and, if so, whether civilians will punish them for it. These expectations are a function of overlap between the preferences of the civilian and the military players, and the political strength of the actors" (Feaver 2003, 3).

It's worth pausing here to consider more closely what Feaver means by the terms "working" and "shirking." "Working" is relatively unproblematic—an agent is working when she is diligently pursuing the tasks assigned to her by her superior. In the case of the military, the military is working when it diligently seeks to fulfil the wishes of its civilian overseers. "Shirking," in this context, means more than simply failing to work. For the military may be vigorously pursuing military and/ or policy goals, but it will still be shirking if those goals do not correspond with the desires of the civilian principal.

"What civilians want" is of course a complex and multidimensional issue in the context of civil–military relations, far more so than in the traditional economic applications of principal–agent theory. Feaver points out that, in structural terms, the desires of the civilian principal can be viewed as two-fold. Firstly, civilians want to be protected from external enemies. Secondly, they want to retain political control over the military, and ensure that it is in fact the civilian principal that is making the key decisions about the military and its activities. Feaver calls the first of these the "functional goal," and the second the "relational goal."

In principal–agent theory terms, the problem of getting the agent to work in the desired manner is called the "moral hazard problem." Feaver points out that in the general literature on principal–agent theory, there are two distinct opinions in this regard. On the one hand there are those who contend that the best way to ensure that the agent is working is by applying the best available monitoring system. On the other hand there are those who believe that the superior approach is to implement measures aimed at adjusting the agent's preferences to increasingly coincide with those of the principal. Feaver's theory draws on insights from both. Importantly, Feaver adds an important additional consideration that is seldom addressed in principal–agent theory, namely "how agent behavior is a function of their expectation that they will be punished if their failure to work is discovered; traditional principal–agent treatments assume punishment is automatic but … I argue … that assumption must be relaxed when analyzing civil–military relations" (Feaver 2003, 56).

Envisaged in this way, civil–military relations are viewed as a game of strategic interaction in which each side attempts to achieve outcomes that maximally promote that side's interests. This is clearly a significantly different approach to that favored by Huntington and Janowitz, where nonmaterial factors such as identity and moral commitments arguably play the central role. These factors are not, however, irrelevant to Agency Theory. Instead, Feaver argues, Agency Theory provides a framework of analysis against which the influence of these factors may be measured and assessed. In particular they can be understood in terms of the attempt to seek convergence between the preferences of the civilian principal and the military agents.

The preference for honor is one of three preferences that the military agent is assumed to hold by "Agency Theory." Another is the preference for specific policy outcomes. This is different to the usual principal–agent relationship, where the agent generally has no interest in which economic policy is pursued by the principal. The military agent, on the other hand, has a preference for policies that do not needlessly risk his life. In addition, the military agent has a preference for policies that give overwhelming supremacy on the battlefield. The last basic military preference is one for maximal autonomy, which translates in large part into a desire for the minimization of civilian interference in military affairs. All of these preferences can lead the military to attempt to influence policy in ways that undermine civilian control. In terms of democratic governance, this is pernicious even when such interference leads to better security arrangements than would otherwise have been achieved. Dealing with this is made all the more difficult by the fact that the military agent carries a particular moral status—her willingness to make the ultimate sacrifice for her country acts in some sense as a moral counterweight to the civilian principal's political competence. As a consequence, "the moral ambiguity of the relationship bolsters the hand of a military agent should he choose to resist civilian direction" (Feaver 2003, 71–2).

The other central problem for civil–military relations, in terms of Feaver's principal–agent derived theory, is the "adverse selection problem." This is the problem facing the principal in selecting which agent to contract with to undertake the required task. The agent has a strong incentive to portray herself as being far more diligent than she is, in order to ensure that she is contracted, which complicates the principal's task of selecting the best possible agent for the job. In civil–military terms the task is one of leadership selection—which potential senior officers are most likely to lead the military to work rather than to shirk? The special nature of the military context gives this problem a unique twist. Feaver seems right to point out, for example, that it is "at least plausible" that the sort of personality that is advantageous on the battlefield is by nature problematic in terms of the principal–agent relationship (Feaver 2003, 72). Indeed, as we have seen, this is one of the central reasons that Huntington stresses a sharp differentiation of the civilian and military spheres.

Civil–Military Relations, Contracting, and Delegation

We turn now to consider whether or not there is any substantial difference in the nature of the essential strategic interaction between civilian principals and state military agents, on the one hand, and that between civilian principals and private military agents on the other. Critics of the private military industry have strongly expressed their concerns in this regard. Herfried Münkler, for example, expresses the opinion that,

> a continuation of this tendency [towards military privatisation] would have enormous political consequences, as the armed force would be subject to weak control by governments (linked only by the employment relationship). … Privatized warfare would rapidly take on a disastrous life of its own, in accordance with the laws of the market. (Münkler 2005, 134–5)

In what follows I shall begin to explore whether such concerns are justified.

A key commonality between civilian principals and state military agents, on the one hand, and civilian principals and private military agents on the other, but one that is not necessarily obvious when civil–military relations is viewed outside of the Agency Theory framework, is that in both cases the relationship is one of *delegation*. This is most obvious in the case of the private military company, for the notion of delegation is built into the very essence of the commercial contract. As Feaver points out, however, the same goes for the state military: "In the civil–military context, the civilian principal contracts with the military agent to develop the ability to use force in defense of the civilian's interests" (Feaver 2003, 57). This is an important point to recognize, for the essential objection made by many opponents of military privatization is that it is inappropriate to delegate military tasks to non-governmental organizations. Peter W. Singer, for example, writes that,

> [w]hen the government delegates out part of its role in national security though the recruitment and maintenance of armed forces, it is abdicating an essential responsibility. When the forms of public protection are hired through private means, the citizens of society do not enjoy security by right of their membership in a state. Rather, it results from the coincidence between the firm's contract parameters, its profitability, and the specific contracting members' interests. Thus, when marketized, security is often not about collective good, but about private means and ends. (Singer 2003, 226)

However, once it is recognized that state military forces are distinct organisations to which elected civilian governments delegate some of the responsibility of protecting the state and pursuing the state's vital interests, it is hard to see how this can stand as a meaningful objection. For delegation is the essence of democracy: citizens delegate to their elected representatives the responsibility to rule, and those representatives in turn delegate to others the specific tasks that must be carried out in order to actualize that rule. These relationships can also be expressed in terms of *contracts*, as in social contract theory. The form of the contract between the civilian principal and the military agent may look somewhat different to more standard contractual arrangements—involving as it does various cultural rituals, symbols and honors—but a contract it is nonetheless.

Feaver points out that "[t]he primary claim of the principal–agent literature is that delegation need not be an abdication of responsibility" (Feaver 2003, 55). This is because delegation need not mean a loss of control. Feaver convincingly argues that a number of means are available to civilian principals by which to make state military agents to do what they are supposed to do. It is my contention that those means are just as applicable to private contractors as they are to military personnel. I cannot argue that here, however, as it is necessary to first establish the more fundamental question, that of whether private agents alter the fundamental strategic relationship that generally holds between military forces and civilian agents. In so doing it is important we keep in mind Avant's point that "There is generally some loss of control, or slippage, associated with *any* delegation; the question should not be how private choices compare with an ideal relationship, but how they compare with other available options" (Avant 2005, 43).

What underlies the strategic relationship between civilians and the state military is the fact that there is a strong likelihood of a divergence of preferences between the two parties. This is the essence of the principal–agent problematique. Although there may at times be shared preferences among both parties, the very nature of the two-sided relationship opens up the potential for divergence. Various factors contribute to this potential—differing views of what national security goals should be, differing views of how to pursue those goals, the natural desire for the military to build the biggest "empire" possible, the natural desire of the civilians to limit the size and scope of the military to what they deem is necessary to achieve security, and so on. As Feaver points out, "the military has the ability and sometimes also the incentive to respond strategically to civilian delegation and control decisions – in the jargon of principal-agency, to shirk rather than to work" (Feaver 2003, 57). This defines the strategic relationship on the military's side. On the other side, the civilian principal has the desire to ensure that the military works rather than shirks, and so sets out to set in place mechanisms for making this so. Feaver sums up the results as follows,

> In sum, civil–military relations is a game of strategic interaction. The "players" are civilian leaders and military agents. Each makes "moves" based on its own preferences for outcomes and its expectations of how the other side is likely to act. The game is influenced by exogenous factors, for instance the intensity of the external threat facing the state made up of the players. The game is also influenced by uncertainties. The civilians cannot be sure that the military will do what they want; the military agents cannot be sure that the civilians will catch and punish them if they misbehave. (Feaver 2003, 58)

Returning to our central concern, we must ask the question of whether this description would read any differently if Feaver had written it about civil–(private) military relations, in which the "players" are civilian leaders and private military agents? Given that "Agency Theory" is derived from principal–agent theory developed in the context of commercial relationships between employers and employees, and that this is essentially the same kind of relationship as applies between civilian leaders and private military agents, there seems very little reason to think that the strategic relationship should be any different. The only significant difference between the state military and private military companies in this context is that there is generally only one official state military for any particular country (albeit one divided into different services), while there is potentially a plethora of private military companies competing for state contracts. If anything, however, this difference favors the private military companies when it comes to the application of Agency Theory. As Feaver points out, there is something anomalous in applying principal–agent theory to the issue of civil–military relations because "[t]here is not really a market of agents; the civilian cannot hire from many different militaries to do its work. The principal can create new military agents, and does so from time to time, but there is something of a monopoly in providing security"[2] (Feaver 2003, 314, note 6). Feaver argues implicitly, and I believe successfully, that this anomaly does not undermine the applicability of the principal–agent framework to the sphere of civil–military relations. But it is worth noting that no such anomaly applies to the relationship between private military companies and state employers. Instead the latter relationship is a classic case of principal–agent interaction.

In broad terms then, it seems that the basic strategic relationship between civilian principals and state military agents, on the one hand, is not significantly different to that between civilian principals and private military agents. In the next section I consider the specific goals civilian principals have in the context of military force, and evaluate whether there is any significant divergence here between what civilians want from state military forces and private military forces.

Functional and Relational Goals

The central challenge presented to civilian principals by the nature of the strategic relationship at the heart of civil–military relations is the danger that military shirking will lead to significantly "suboptimal arrangements" ranging from battlefield collapse, unwanted wars and coups; to simply placing an unwarranted economic burden on society. This problem is minimized when there is a convergence of

---

2   In the same footnote Feaver notes that in traditional civil–military relations "the government enjoys a monopsomy in purchasing security." It is generally held that, in terms of social contract theory, it is essential that the state hold the monopoly on violence. This is a point that is sometimes raised to argue for the illegitimacy of private military companies. It's a nice question, however, whether it is not more accurate to say that social contract theory requires the state to have a monopsomy with regard to violence. Unfortunately the constraints of this chapter do not allow me to pursue this question, though it is my hunch that this is indeed what is, in fact, implied by social contract theory. The undermining of this monopsomy seems to me far more of a threat to appropriate civil–military relations than is the loss of the state military's monopoly on force.

preferences between civilians and their military agents. This can sometimes be achieved through, for example, promoting senior officers who have shown themselves to share the preferences of their civilian masters. Feaver, however, argues that there are limits to how far this goes:

> For starters, military communities have strong identities that mark them as "different" from those of civilians, and this is deliberately cultivated and signified through uniforms, oaths of office, rituals, and so on; there is, in other words, some irreducible difference between military and civilian, and this will naturally extend to different perspectives. Moreover, the civil–military difference is compounded by the different role each plays, one as principal, the other as agent; there is a *de minimis* difference in perspective that attends agency, hiring someone else to do something for you. (Feaver 2003, 60)

The range of monitoring and punishment mechanisms Feaver outlines are designed to address this unavoidable problem in civil–military relations. But a prior question to that of means of securing compliance is that of just what goals civilian principals have with which they desire military agents to comply. As we have seen, Feaver singles out two central goals that it can be presumed that civilian principals have with regard to military servants: "Civilians want protection from external enemies *and* want to remain in political control over their destiny" (Feaver 2003, 62). Feaver calls the first of these goals the functional goal, and the second the relational goal. These goals can be further broken down into specific tasks:

The functional goal includes the following:

1. Whether the military is doing what civilians asked it to do, to include instances where civilians have expressed a preference on both the "what" and the "how" of any given action;
2. Whether the military is working to the fullest extent of its duty to do what the civilians asked it to do;
3. Whether the military is competent (measured by some reasonableness standard) to do what civilians asked it to do.

The relational goal can be broken down into the following:

1. whether the civilian is the one who is making key policy decisions (i.e., no *de facto* or *de jure* coup) and whether those decisions are substantive rather than nominal;
2. whether the civilian is the one who decides which decisions civilians should make and which decisions can be left to the military;
3. whether the military is avoiding any behavior that undermines civilian supremacy in the long run even if it is fulfilling civilian functional orders (Feaver 2003, 62).

Returning to our central quest, we must ask at this point whether there is anything different here if we substitute private military companies for state military forces? Once again it is hard to see why there should be any difference.

While the desires civilian principals have for the behaviour of private military agents appear to be little different from the desires they have for the behaviour of state military agents, there are clear differences between the preferences of state militaries and those of private military forces. One important difference here between state militaries and private forces arises from the fact that private military companies are only paid when they are deployed, while state militaries are paid even when they are not employed in their primary warfighting role. In terms of the strategic game described by Agency Theory, the ideal situation for the state military is where civilians view the strategic environment as a threatening one and fund the military accordingly, but where the military does not in fact have to deploy or fight, thereby avoiding all the costs incurred. The military agent has a preference for policies that do not needlessly risk his life, as well as for policies that give overwhelming supremacy on the battlefield. Feaver points out that this results in a danger that state militaries will use their advisory role to pump up estimates of what military power is required to resist possible threats, while also using their advisory role to minimize the number and extent of their actual deployments. As Feaver puts it, "[T]here is an exceedingly blurry line between advising against a course of action and resisting civilian efforts to pursue that course of action. Sometimes negative advice can rise to the level of shirking, especially if the advice is exaggerated" (Feaver 2003, 62).

Following the same strategic logic, it appears that private military companies will be inclined to shirk in the other direction. As they are paid to deploy and receive no income from the state when not deployed, they are likely to be prone to downplaying the likely cost of intervention while at the same time exaggerating the benefits thereof. Thus, where the state military is strategically inclined to shirk in the direction of inertia, the private military company is inclined to shirk by seeking to deploy more often than is in fact necessary. Some commentators express concern of what they see as the broader implications of this. Musah and Fayemi, for example, argue that "it is in the interest of the new mercenaries that the world remains in a perpetual state of instability" and that, as a result "their 'solutions' are often short-term" (Musah and Fayemi 2000, 28).

Does this difference amount to a significant problem when considered from the perspective of the civilian principal? I argue that it is not. The first point to note is that both preferences, if carried through, result in shirking, and so there is no difference in the *type* of problem these preferences potentially raise. In both cases the civilian principal has a duty to make herself aware of these preferences and their potential dangers, and to act to ensure that those potential dangers do not become actual. Thus, for example, a civilian principal who is aware of the private military company's potential to exaggerate the benefits of military action can employ another company to act purely as an advisor on these matters—i.e. the latter company will gain no benefit if a decision is made to deploy. The fact that the private provision of military services is competitive and involves numerous players in a market offers the civilian principal greater flexibility of this kind than when dealing with the monopoly agent that is the state military force.

A second relevant point here is that, given that in a democracy civilians "have the right to be wrong," there should in a democratic society be a preference for agents that will be more responsive to civilian directives. Given that, as we have seen, the strategic preference of the state military is well-paid inactivity, while the strategic preference of the private military company is active employment, there seems to be at least a small reason to prefer the private agent in this regard.

A related point that favors the use of private military agents by civilian principals is the fact that private military companies have no special societal status, unlike their state military counterparts. The state military agent has a unique moral status in society, as a result of her willingness to make the ultimate sacrifice for her country. This can give the state military agent leverage should she seek to resist civilian direction. In Feaver's words, "the moral ambiguity of the relationship bolsters the hand of a military agent should he choose to resist civilian direction" (Feaver 2003, 71–2). The private military company has no such moral status, despite potentially carrying out precisely the same missions and carrying precisely the same level of risk as the state military. Once again it seems that the private military agent is, at least in principle, likely to be more responsive to civilian direction than the state military agent.

As mentioned above, there are two further preferences that the state military agent is assumed to hold by Agency Theory. The first is the preference for honor. I will return to this issue at the end of this chapter. It is however worth making one point here on this issue. As Feaver points out, "[h]onor permeates the famous concept of small-group cohesion, the factor that makes human beings willing to risk their lives" (Feaver 2003, 73). Numerous authors argue that the heart of small-group cohesion in military forces is commitment to one's buddies in the group, rather than commitment to broader ideals.[3] If this is indeed true, then there is little reason to think that private units will by their nature lack the cohesion necessary for battlefield success.

The final essential preference displayed by the state military is the preference for maximal autonomy. In Agency Theory terms, this autonomy acts as a substitute for profit sharing. Autonomy is what the state military receives in exchange for obedience to the civilian principal – this is the heart of Huntington's notion of "objective control." But Kohn points out that "while 'objective' civilian control might minimize military involvement in politics, it also decreases civilian control over military affairs" (Kohn 1997, 143). The same does not go for the private military company, where it is profit rather than autonomy that is the key preference. While this does not mean that autonomy is not a value for the private military agent, there is good reason to expect that the private military agent will be willing

---

3   See for example the account of the complexities of battle-motivation in Chapter 7 of Richard Holmes' classic work, *Firing Line* (1985).

to trade autonomy for profit, thereby potentially increasing civilian control. Once again, therefore, it seems that the private military company looks somewhat better than the state military agent from the perspective of a civilian principal who is aware of "Agency Theory."

Information Asymmetries, Adverse Selection and Moral Hazard in Civil–Military Relations

> Principal–agent relationships involve information asymmetries. Both sides share common information; in the civil–military context, they know who the domestic players are, the size of the defense budget, the general identity and nature of their enemies. They also share a common history and political memory. But each has private information that is discerned only dimly by the other. (Feaver 2003, 69)

In the case of civil–military relations, the private information held by state military forces includes expert knowledge on issues like weapons system capabilities, tactics, logistics and morale, as well as inside knowledge regarding the general attitude within the military towards the directives of the civilian principal. For civilians, the private information includes insight into political realities and preferences. Overall, however, as Feaver points out, "information asymmetries favor the [state] military agent" (Feaver 2003, 69). This is particularly so when the state military is deployed and engaged in combat operations – the very nature of distant and chaotic engagements makes it extremely difficult for civilians to monitor the military.

Is the information asymmetry between state military forces and civilian principals matched by a similar asymmetry between private military companies and their state employers? Certainly, at the most basic level, the question must be answered in the affirmative. As Feaver implies in the quote at the beginning of this section, the very nature of the principal–agent relationship ensures the existence of some informational asymmetry. But there are significant differences between the private military company and the state military force that suggest that the asymmetry might be less pronounced in the case of the private military company. For one thing, as discussed in the previous section, the substitution of autonomy for profit in the case of the state military force increases the likelihood that the state military force will be more resistant to civilian monitoring than will the private military company. For another thing, the nature of the market for the private provision of force increases the incentive for private companies to seek to earn the trust (and therefore the contracts) of the civilian principals by making themselves as open to the civilians as possible. Christopher Kinsey, for example, argues that the future success of private military companies will be determined more on how much "corporate social responsibility" they display than on their ability to find new markets (Kinsey 2008). Avant agrees: "Conceptions of proper behavior, such as the codes of conduct and standards in vogue among advocates of corporate social responsibility, can be important in setting expectations and norms within which the market works" (Avant 2005, 220).

Apart from information asymmetries, principal–agent interactions in general, and civil–military relations in particular, are also afflicted by the adverse selection problem and moral hazard. As Feaver explains,

> Adverse selection refers to the moment of hiring in the employer metaphor. Has the employer hired someone who is naturally a hard worker or has he been deceived by the interview and hired a lout? Just how closely aligned are the preferences of the agent and the principal? The adverse selection problem means, in the first instance, that the employer cannot know for certain about the true preferences and capabilities of the applicant. But adverse selection is more than mere uncertainty about the applicant. It also refers to the fact that the very act of hiring creates perverse incentives for the agent to misrepresent himself, which thereby increases the chances that the principal will hire a lout: it is hard to verify the true type, and the lout has a great incentive to appear even more attractive than a good worker. ... More generally, adverse selection can extend beyond the hiring phase to include all those situations in which the agent presents himself, or some proposal, to the principal for approval or decision. For instance, it means that because of their informational advantage over superiors, subordinates tend to propose policies that benefit their own interests rather than the interests of the superiors. (Feaver 2003, 72–3)

While adverse selection does not confront the traditional civil–military relationship in a direct way, given that the civilian principal is not faced with a choice as to which agent to employ, Feaver points out that in this context the problem appears when civilians decide on which military officers to promote to senior rank. There are particular difficulties here because the personality of a person who is likely to succeed on the battlefield is not one that succumbs comfortably to oversight by civilians who, in military matters at least, are in all likelihood her inferiors. As Feaver observes, "[o]ne of the major concerns of traditional civil–military relations theory was precisely the great divergence of viewpoint between what Huntington called the liberal civilian ideology and the military mind" (Feaver 2003, 73).

The other area where adverse selection appears in the traditional civil–military relationship is in the budget process, in which the civilian "selects" what warfighting capability it will pay for, on the basis of proposals put forward by state military organizations.

> Again, because the military has an information advantage it can advance artfully drawn proposals that appear to meet civilian needs but in reality are tailored to its own interests. In the extreme, adverse selection might lead civilians to adopt policies they think will increase the military's ability to protect society

but that in fact will increase the ability or even the propensity of the military to undermine society (Feaver 2003, 74).

Peter W. Singer, for one, thinks that the adverse selection problem is particularly problematic for states that employ private military companies,

> This issue of adverse selection becomes particularly worrisome when placed in the context of the industry, with its layers of moral hazard and diffused responsibilities. Thus, even if Private Military Firms are scrupulous in screening out their hires for human rights violations (which is difficult for a firm to accomplish, given that most of its prospective employees' resumes do not have an 'atrocities committed' section), it is still difficult for them to monitor their troops in the field completely. (Singer 2003, 222)

I cannot examine the specifics of monitoring in this chapter, though I expect to do so in a forthcoming work. For our purposes here the interesting question is whether the fact that adverse selection is only indirectly applicable to traditional civil–(state) military relations (because of the state military's monopoly on the provision of military forces) is something that shows that Agency Theory's applicability to private military companies is limited?

Once again the obvious rejoinder is that it is the relationship between the private military company and its civilian state employer that is the paradigm case of the principal–agent relationship, and it is the traditional civil–(state) military relationship that must be manoeuvred somewhat to fit this framework. The adverse selection problem very clearly applies when the state is choosing which private military company to employ, in exactly the same way as it applies when the state is choosing which contractor of any type to employ. This problem is exactly that, a problem. But as Feaver makes clear it is a problem that applies in the context of state military forces as well, and it is one that in that context can be addressed by some or all of the mechanisms he outlines. It remains to be seen whether the same or similar endeavours on the part of the civilian principal will successfully address the problem as it appears in the civil–(private) military context, though it is my view that there is no fundamental reason why they should not.

I come finally, and most briefly, to moral hazard.

> Moral hazard refers to the behavior of the employee once hired. Like adverse selection, moral hazard refers at a general level to the problem that principals cannot completely observe the true behavior of the agent and so cannot be certain whether the agent is working or shirking. It has an additional specialized meaning based on the perverse incentives in the agency relationship. Employees have an incentive to shirk rather than work; if you can get paid for doing less, why do more? The principal, of course, tries to minimize shirking because it is inefficient (Feaver 2003, 74).

Moral hazard afflicts the civil– (state) military relationship in a more direct way than adverse selection. Because moral hazard is structurally very similar to the adverse selection problem, albeit applied downstream, it seems clear that the comments made above apply equally well here, so I will not belabour the point by repeating them.

In sum, what is missed by many critics of private military companies is that their concern over the impact of military privatisation on civil–military relations ignores the fact that civil–military relations are by their very nature fraught. Feaver's analysis leads to the expectation that "this principal–agent relationship [i.e. that between civilian principals and state military forces] should be particularly characterized by distrust and friction, and any equilibria of delegation and control are unlikely to endure, giving way to new arrangements as costs and benefits shift" (Feaver 2003, 72). As we have seen it appears that not only do private military companies not fundamentally part company with state military forces over the nature of their relationship with civilian principals, but on some counts private military companies also fare slightly better (from the perspective of the civilian principal) within that relationship, from a broad conceptual point of view at least. The one important exception to this, which I have deliberately set aside until now, relates to the role that honor plays in the relationship between state military forces and the states they serve. It is to that I now turn.

Honor and Allegiance

> When can their glory fade?
> O the wild charge they made!
> All the world wonder'd.
> Honor the charge they made!
> Honor the Light Brigade,
> Noble six hundred.
>> Alfred Lord Tennyson, *The Charge of the Light Brigade*

The reader might rightly complain that the discussion of strategic relationship between elected principals and their private or state military servants that has taken up the bulk of this chapter misses the point in one important respect. The chapter has been addressing the issue of *control*, but the topic at hand is *allegiance*, and the latter concept is not reducible to the former. I concede this point without hesitation, and propose that at least part of what accounts for the gap between control and allegiance is the idea of *honor*.

As we saw above, Feaver points out that one of the primary preferences of state military agents is the preference for honor. Honor is a notoriously difficult concept to pin down,[4] and I will not attempt to derive the definitive definition here.

---

4   See, for example, the interesting papers that address this topic in Volume 4:3 of the *Journal of Military Ethics*, 2005.

It will be sufficient for the purposes of this chapter to accept the uncontroversial claim that, as Peter Olsthoorn put it, "honor is a motive for making sacrifices" (Olsthoorn 2005). Honor is, of course, not the only motive for making sacrifices – love is another – but it is without question one of the main reasons that individual members and units of state military forces are prepared to cross into the badlands that lie between manageable risk and certain death.

This preference for honor, and the consequent willingness to make sacrifices (including the ultimate sacrifice) must be acknowledged as a significant difference between state military forces and private military companies. That is not for a moment to suggest that individual members of private military or security companies cannot display honor, or make heroic sacrifices. The point is rather that the companies themselves, and the managers that "command" them, cannot (indeed *ought* not) take the preference for honor as a guiding principle for their actions, at least not to the extent that the quest for honor demands sacrifice.[5] There are circumstances in which the military commander can, and should, send or lead her unit on missions in which there is a significant likelihood that successfully completing those missions will require the death or injury of some, many, or all of those under her command. This is because there is built into the idea of national military service a presumption that what would be supererogatory for the average citizen—the willingness to give up life and limb—is, under appropriate circumstances, normative for military personnel. The same is certainly not true for the private military manager. While individuals who join private military companies accept a significantly higher level of risk in their field of employment than in most other occupations, the private military manager has no right to expect those under his "command" to sacrifice themselves for some higher good. Sacrifice has no place in the cost-benefit analysis that is at the heart of commercial soldiering.

What is clear from this is that while there is a significant range of operations in which the "allegiance" of private military managers and military commanders can be viewed as strategically equivalent and on a moral par, there is also a range of operations, those that may meaningfully require "allegiance unto death," for which private forces would be both morally and practically inappropriate.

There is, however, a flip-side to this. The central reason that it is appropriate to deploy national military forces in circumstances where sacrifice might be demanded is that these forces owe a duty to the state, a duty that sometimes extends beyond "manageable risk." But that duty does not apply in cases of armed humanitarian intervention where the sole or primary driving force behind the intervention is the international community's "responsibility to protect." This leads to an awkward paradox. While states may arguably have an obligation to use their military forces for the sake of humanity, there is no commensurate obligation on individual military personnel to risk their lives for humanity. As Michael Gross puts it,

---

5   It might be helpful here to differentiate between *respect* and *honor*. While it is certainly appropriate (and good business practice) for private companies to seek to earn respect, it would not be appropriate for them to seek honor.

The State/Private Military Continuum

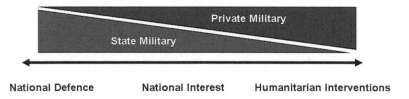

Figure 10.1   The State/Private Military Continuum

One may reasonably ask whether American soldiers, for example, volunteer for humanitarian duty when they swear to "support and defend the Constitution of the United States against all enemies, foreign and domestic." There is nothing here to suggest that soldiers agree to fight against foreign armies that do not threaten American Security. Without volunteers who consent specifically to humanitarian duties, states have no choice but to refuse their international obligations. (Gross 2008, 219)

Gross argues that this paradox can be avoided by turning to large conscript armies as a means to mitigate the risk carried by each individual soldier and thereby bringing the level of risk to within acceptable bounds, thereby legitimating the deployment of these soldiers in support of humanitarian ideals. Gross therefore contends that, contrary to current practice, "A conscript army is ... a first, and not a last resort measure that states should turn to as they consider wars of humanitarian intervention" (Gross 2008, 225). Whatever one thinks of Gross' interesting argument, it is surely evident that there is in fact a step that must be preferred to the deployment of *any sort* of national military force for humanitarian interventions, namely the deployment of private military companies. For while national military personnel clearly do not "sign up" to risk their lives for humanity, this is exactly what private forces can voluntarily sign up to do, and provided the cost–risk benefit is in their favor, this is what they can be relied upon to do.

In sum, then, we might say that the morally appropriate deployment of state and private military forces fall along a continuum. At one end there are operations of national emergency, in which it might well be appropriate to expect members of national military forces to be prepared to lay down their lives for their countries. At this end of the spectrum honor and political obligation imposes on state military commanders and their subordinates a duty of "allegiance unto death" which cannot be held to apply to private military managers and contractors. It would therefore be morally inappropriate, and practically unwise, to deploy private military forces for such operations. At the other end of the spectrum are military operations in support of humanity. Because state military commanders and their subordinates do not individually owe allegiance to humanity, but private military contractors may voluntarily choose to contractually bind themselves in service of humanity, there is a moral duty by states to prefer private military companies over state military forces for such operations. Between these extremes there lies a range of operations which might reasonably be carried out by either state or private military forces, depending on the degree of national interest and the degree of risk involved. In this middle range civilian control of military forces, whether state or private, is generally an adequate alternative to honor-driven allegiance. As we have seen Agency Theory provides a useful tool by which to understand this control, and private military companies and their managers do not appear to fare any worse than state military forces and their commanders under this framework.

References

Avant, D.D. (2005), *The Market for Force: The Consequences of Privatizing Security* (Cambridge: Cambridge University Press).

Baker, D-P. (2007), "Agency Theory: A New Model of Civil–Military Relations for Africa?," *African Journal on Conflict Resolution*, 7:1, 113–36.

Feaver, P.D. (2003), *Armed Servants: Agency, Oversight, and Civil–Military Relations* (Cambridge, MA: Harvard University Press).

Gross, M.L. (2008), "Is there a Duty to Die for Humanity?: Humanitarian Intervention, Military Service and Political Obligation," *Public Affairs Quarterly*, 22:3, July 2008, 213–29.

Kinsey, C. (2008), "Private Military and Security Companies and Corporate Social Responsibility," in Alexandra, A., Baker, D-P. and Caparini, M. (eds), *Private Military and Security Companies: Ethics, Policies and Civil–Military Relations* (London: Routledge) 70–86.

Kohn, R.H. (1997), "How Democracies Control the Military," *Journal of Democracy*, 8:4, 140–53.

Krahmann, E. (2008), "The New Model Soldier and Civil–Military Relations," in Alexandra, A., Baker, D-P. and Caparini, M. (eds) *Private Military and Security Companies: Ethics, Policies and Civil–Military Relations* (London: Routledge) 247–65.

Münkler, H. (2005), *The New Wars* (Cambridge: Polity Press).

Musah, A-F. and F.J. 'Kayode (eds) (2000), *Mercenaries: An African Security Dilemma* (London: Pluto Press).

Olsthoorn, P. (2005), "Honor as a Motive for Making Sacrifices," *Journal of Military Ethics*, 4:3, 183–97.

Singer, P.W. (2003), *Corporate Warriors: The Rise of the Privatized Military Industry* (Ithaca, NY: Cornell University Press).

PART V
Combat Behavior and Training

Chapter 11

Deconstructing the Evil Zone: How Ordinary Individuals Can Commit Atrocities

Paolo Tripodi

[W]e thought that perhaps the units had included an unusual number of men of inferior quality. When we thought only Charlie Company had been involved in the incident [the My Lai massacre], we had requested the deputy chief of staff for personnel to make an analysis of the men in that company. The result was a fact sheet that in the main concluded that the men of Charlie Company were about average as compared with other units of the Army.

U.S. Army Lieutenant General W.R. Peers (Peers, 1979, 231)

Over the past few decades, scholars from several disciplines have investigated why and how atrocities such as the My Lai massacre occur (Browning 1993, Osiel 2009, Newman and Erber 2002, Hinton 2005). Indeed, additional knowledge is crucial to develop instruments that can prevent these atrocities from happening in the future. Therefore, understanding atrocities is an effort that goes beyond an important intellectual and academic exercise. This chapter explores some of the circumstances that may increase the likelihood that members of a military unit might commit an atrocity, the focus is on "atrocities by situation" (Tripodi, 2010). This term refers to atrocities that might be better explained through an understanding of the situation individuals operated in, rather than by the individuals' disposition.

The premise of this chapter is that perpetrators of "atrocities by situation" are evil doers, but they might not necessarily be evil persons. Indeed, when an atrocity by situation is perpetrated by a military unit, a number of individuals might be evil persons, but the large majority of them likely are not. Therefore, explanation for their actions cannot be a result only of these individuals' "disposition;" the exploration of why atrocities happen should go beyond the individual. U.S. Army Brigadier General H.R. McMaster, in "Ethics Education for Irregular Warfare," stressed that, "Irregular warfare creates abnormal particular pressures on those who participate in it. Some respond to these pressures by acting at variance to professed military values" (McMaster, 2009, 15).

This chapter focuses on a specific area of atrocities by situation, it will explore a space that could be defined as the "evil zone," a space in which ordinary individuals might perpetrate an atrocity or commit evil acts. Indeed, the exploration of the conditions that might lead individuals to step into the evil zone constitutes one element that can help us understand individuals' behavior. It should be acknowledged that soldiers who are responsible for an "atrocity by situation" might have gone through an extremely aggressive training program; they might have developed a sense of cohesion that, instead of acting as a positive aggregating factor for all members of the unit, encourages dangerous, devastating group dynamics (Lifton 1998).

Thus, this chapter explores the physical and psychological dimension of the battlefield and its impact on the individual. Such an exploration uses the experience of soldiers who have been involved in a variety of operational settings. The chapter relies on personal narrative of soldiers and authors who spent a significant amount of time attached to military units in a combat zone. Indeed, personal narrative is subjective and at time might be subject to the authors' personal biases and the way memory might distort important events related to their experience. However, personal narratives offer insight into the environment in which these individuals were operating. Narratives also provide an opportunity to identify important common features that will enhance our ability to analyze soldiers' behavior.

Disposition, Situation and Evil

In August 2009, at a speech at a club in Columbus Georgia, a man in his mid-sixties said "There is not a day that goes by that I do not feel remorse for what happened that day in My Lai. I feel remorse for the Vietnamese who were killed, for their families, for the American soldiers involved and their families. I am very sorry."[1] More than forty years after Calley led his platoon to massacre defenseless civilians as a young Army Lieutenant, he is still haunted by the event. Even more dramatic was the testimony of another participant in the massacre, Private Varnardo Simpson. He explained that on the day of the massacre he "lost all sense of direction, of purpose. I just started killing in any kind of way I could kill. It just came, I didn't know I had it in me … after I killed the child my whole mind just went, it just went" (*Remember My Lai*, 1989).

Scholars from a variety of disciplines have warned us about our mistaken belief that character, even a strong solid character, does not constitute a reliable protection against evil behavior. Jonathan Shay, a psychiatrist who undertook extensive research on Vietnam veterans emphasized that we are strongly inclined to believe that "good character stands reliably between good persons and the possibility of horrible acts" (Shay, 1995, 31). Owen Flanagan provided an explanation of this mistaken perception. Elaborating on Lee Ross's fundamental attribution error,[2] Flanagan wrote that such an approach "involves an inclination to overestimate the impact of dispositional factors [the individual traits] and to underestimate situational ones" (Flanagan, 1991, 306).

---

1   http://www.pbs.org/wgbh/americanexperience/features/timeline/mylai-massacre/2/.

2   According to Ross and Nisbett "People's inflated belief in the importance of personality traits and dispositions, together with their failure to recognize the importance of situational factors in affecting behavior, has been termed the "fundamental attribution error" (Ross, Nisbett, 1991, 4).

In analyzing the behavior of individuals who are responsible for an evil deed, often much attention is placed on the individuals' behavior while the situation— how it evolved, for how long, and other key circumstances—is neglected. The attempt to provide an explanation for a given action focuses on the individual and his or her character. Thus, the role and power of other factors that may influence an individual's action is underestimated or not even considered. We tend to believe that a man who would do what is morally right or "normal" under normal circumstances would do what we consider to be morally right and normal under abnormal conditions.

Philip Zimbardo has rightly pointed out that "we want to believe in the essential, unchanging goodness of people, in their power to resist external pressures, in their rational appraisal and then rejection of situational temptations. We invest human nature with God-like qualities" (Zimbardo, 2008, 211). Socio-psychology has shown evidence that in many cases evil behavior is not explainable solely through an exploration of the personality of the individual, but is also dependent upon where that individual was when he committed an evil act. According to James Waller "Mainstream social psychology has long believed that what really matters is not who you are, but where you are. Decades of research have hammered home the power of the situation in influencing our thoughts, feelings and behaviors" (Waller, 2007, 230).

For the last five years, I have discussed applied ethics with many officers at the U.S. Marine Corps University in Quantico, VA. In one of my lectures, I present a false moral dilemma. I ask the officers to think about the course of action they would select, but also to pay attention to the amount of time they need to make a decision. The following is an abridged version of the example I present to the class:

> You have been selected for a two week embassy duty (they chose whatever embassy they like). For your expenses, a good friend unrelated to your profession opens a bank account for you, which allows you to easily move money in and out of the country. At the end of the first week after a busy work day, on the way from the embassy to the hotel where you stay, you find a leather bag containing a wallet with an ID and a few dollars.

Then, I ask the students to decide whether they would drop the bag at the local police station; give it to the concierge at the hotel and ask him to contact the owner; call the owner directly; or, get rid of the bag and the ID and keep the money. I tell them to focus on how long it has taken them to make one of the decisions.

In this unscientific exercise, nearly 100 percent of the audience promptly selects the "right" course of action: Give the money back to the owner. Indeed, the first three choices are just different possibilities for what is essentially just one course of action: give the money back. I continue to present them with the same scenario a few more times and progressively increase the amount of money they find in the bag up to a few million dollars.

Finally, I ask them to think about what they would do if, a few days earlier, they learned that their spouse lost his or her job. Then, I introduce another circumstance: Their house is about to be foreclosed, forcing them to use a rather large portion of the savings put aside for their children's education. These elements target the students' emotional side, making the decision-making process considerably less rational, more confused, and significantly slower. I do not ask them to share their answers with the overall student population, but I do ask them to think carefully about how much time it took them to make a decision.

There are situations in which we might be tempted to choose a course of action that we know is wrong. When decision-making is delayed and other unethical options are weighted, we might be seriously distracted from doing what is right. Thus, we might give in to a course of action that would not necessarily be our first choice under normal circumstances. In other words, the abnormal circumstances "tempt" us to deviate from the right course of action. Flanagan noted that "[a] just character has no resources to 'counterbalance' the temptation afforded by knowing that one can pursue self-advantage with impunity" (Flanagan, 1991, 258).

The battlefield is a particularly insidious and highly emotional environment, and it is even more so when there are large numbers of unsympathetic and often hostile civilians. This may be the most difficult environment, even for an individual of solid character, to navigate. In this setting—as well as in peace operations— men and women in uniform live in a situation in which they might have to use force to kill the enemy, a fellow human being. At the same time, soldiers might be under the constant fear of being killed. In many cases, they might experience the loss of fellow soldiers.

There is an extensive literature that deals with the psychological complexities of warfighting and peacekeeping and the huge stressors ordinary men and women have to deal with. (Britt and Adler, 2003; Langholtz, 1998) For the objective of this chapter, it should be noted that individuals acting in such an environment are asked to make decisions in a fundamentally abnormal situation. Therefore, it should be acknowledged that individuals deployed in an abnormal situation might act in an abnormal way, even though such a behavior should not be condoned. This does not necessarily mean that the environment is changing the individual and his or her character, but the environment does have the potential to create great confusion. The value of acknowledging such a possibility should encourage the development of a proper level of mental and moral awareness about the power of situational forces and their impact on the individual deployed in the battlefield.

Although this chapter does not deal with evil *per se*, but rather with an exploration of the situation that can create the conditions for evil to happen, a definition of evil is necessary. Claudia Card's definition of evil is particularly appropriate for this chapter, as she connects evil with atrocities. According to Card, evils "have two basic components: (intolerable) harm and (culpable) wrongdoing" (Card, 2002, 4). In her book *The Atrocity Paradigm. A Theory of Evil*, Card provided an elegant intellectual argument for what she defined as the "atrocity theory" which, in her view is "intermediate" between utilitarianism and stoicism (Card, 2002, 4; 50–72).

The Evil Zone: The Physical Environment

In order to understand how normal individuals might step into the evil zone, it is important to identify as many factors as possible that might play a significant role in individuals' decision making process. Zimbardo stressed that good individuals, to whom he refers to as "good apples," can do evil things if they are placed in a "bad barrel." By deconstructing the evil zone, this chapter investigates what constitutes a "bad barrel," particularly for those operating in a warfighting environment.

The "evil zone" is located in a space in which several forces—sociological, psychological and emotional—interact. It is far removed from what an individual is used to, and would consider "normal." In order to prepare for the challenges of the battlefield, soldiers go through realistic training programs; however, even extremely well trained individuals will find operating and acting in such a complex and ambiguous space very demanding. Very likely, when operating in such a stressful space, chances that individuals might step into the "evil zone" increase.

The individual might perceive the battlefield environment as surreal, alien, and extremely upsetting to all senses. The following are insightful passages from personal narratives of individuals who operated in different situations, ranging from conventional warfare to counterinsurgency.

In October 2006, U.S. Army Lieutenant Shannon Meehan deployed with his tank platoon in Baqubah, the capital of the Diyala province, 30 miles north of Baghdad. The level of violence in the province was very high. Since the beginning of the year, more than 1,000 civilians and 500 government officials had been murdered (Meehan, 2009, 29).

The base where the soldiers were stationed was inhospitable, a place where they spent most of their time on patrol rather than inside the compound. There were few amenities and the temperature was extremely hot (Meehan, 2009, 41–42). It is very likely that Lt. Meehan and his troops had the expectation that the operating conditions in Iraq would have been rough and dangerous. Yet, even with such expectation, they were shocked by the sight they were about to see. A few days after their arrival, Lt Meehan and his troops were returning from an operation to the Forward Operating Base (FOB). According to Lt Meehan:

> The air was dry and the temperature had already risen past 100 degrees. We rolled back towards the base along the main road connecting Baqubah to the rest of Diyala Province, and just outside the city limits, al-Qaida had left a gruesome sign. Decapitated heads lined the highway. They were the heads of Iraqi Police, national policemen whom al-Qaida killed for working alongside the Coalition forces. Flies filled the air around them. (Meehan, 2009, 38)

For the soldiers of U.S. Army Battalion 2-16 who arrived in Iraq in 2007, the situation looked grim from the very beginning. Their first contact with the physical environment of a war zone was in the FOB, Rustamiyah. David Finkel, a journalist who traveled with the battalion, described Rustamiyah as a place in which everything was the color of dirt and stank. "If the wind came from the east, the smell was of raw sewage, and if the wind came from the west, the smell was of burning trash. In Rustamiyah, the wind never came from the north or the south" (Finkel, 2009, 16–17). When the battalion began operating on a number of command outposts, Bravo Company was assigned the area of Kamaliyah.

Initially, the company decided to use an old abandoned spaghetti factory as a base. The Company Commander and the Battalion XO assessed whether the factory would host 120 men. In Finkel's words "it was such a mess" (Finkel, 2009, 48). Not only was the building in poor condition, it also contained a body that was abandoned in the factory water septic tank. "The body, floating, was in a billowing, once-white shirt. The toes were gone. The fingers were gone. The head, separated and floating next to the body, had a gunshot hole in the face" (Finkel, 2009, 46).

The battalion leadership could not find a way to justify hiring and paying someone to remove the body. However, even if they did not have any administrative restriction, they could not find anyone who would remove the body anyway. Clearly, the body had to be removed, as it was unacceptable for the company to move in and share their compound with "Bob," as the soldiers began calling the body. The insurgents helped taken care of the issue. A few days after the battalion had identified the spaghetti factory as a possible base, the insurgents went in and detonated several explosives. The factory was gone and so was "Bob."

Former Marine Corps Captain Tyler Boudreau deployed in Iraq in 2004. With his Marines, he arrived in a small town called Mahmudiyah, south of Baghdad. The building that became the home of the Marines was an old chicken factory, which was "home to many feral dogs as well." The Marines took care of the feral dogs. Yet, Boudreau wrote "in the night, within the sandbagged bunkers, there were rats. They were everywhere. They would crawl over our bodies, under our cots, and inside our packs. We could hear the constant rustling in the darkness as they devoured all those nice snacks that the folks back home had sent to us. So we declared war. On the rats" (Boudreau, 2008, 82).

The stress caused by operating in the physical environment of a conventional war was described effectively by E.B. Sledge, a veteran of the war in the Pacific. Sledge was a young man when he decided to enlist in the Marine Corps in the early 1940s. "Sledgehammer" provided a vivid description of that experience in *With the Old Breed*. The challenging and often terrible experiences he faced were broad and covered aspects of a different nature.

In relation to the battlefield physical environment, Sledge noted:

> It is difficult to convey to anyone who has not experienced it the ghastly horror of having your sense of smell saturated constantly with the putrid odor of rotting human flesh day after day. ...This was something the men of an infantry

battalion got a horrific dose of during a long protracted battle such as Peleliu. In the tropics, the dead became bloated and gave off a terrific stench within a few hours after death. (Sledge, 2010, 142)

United States Army Captain Jaime Perez who served in Iraq as a mental health professional shared Sledge's perception. He noted that "More powerful than visual is smell. People that smell corpses, it is something that stays there, it's so powerful … It's like it's impregnated and you can't get rid of it." (E.F. Tripp, 2008, 196).

In addition to the "awful stench of the dead" Sledge remembers the "repulsive odor of human excrement" and "the odor of thousands of thousands of rotting, discarded Japanese and American rations" (Sledge, 2010, 143). What he saw was as terrible. Sledge stated, "I became quite familiar with the sight of some particular enemy corpse, as if it were a landmark. It was gruesome to see the stages of decay proceed from just killed, to bloated, to maggot-infested rotting, to partially exposed bones – like some biological clock marking the inexorable passage of time" (Sledge, 2010, 143).

Philip Caputo, a Marine Corps Lieutenant and a Vietnam veteran was shocked by the devastating effects of modern weapons on the human body. The young Marine Lieutenant remembers that "we were sickened by the torn flesh, the viscera and splattered brains" (Caputo, 1996, 128). Yet, although both the smell and sight of decaying corpses were extremely disturbing, the sight of killed comrades was emotionally devastating. Sledge described the bodies of dead marines who had been horribly mutilated by the Japanese. One of the dead Marines had been decapitated. "His head lay on his chest, his hands had been severed from his wrists and also lay on his chest near his chin. In disbelief I stared at the face as I realized that the Japanese had cut off the dead Marine's penis and stuffed it into his mouth." Another corpse was "chopped up like a carcass torn by some predatory animal." The Marine noted that "my emotions solidified into rage and hatred for the Japanese beyond anything I ever had experienced" (Sledge, 210, 148).

The battlefield environment is filled not only with gruesome sights, but also with disturbing sounds. Robert Leckie, another veteran of the Pacific war and the author of the compelling book, *Helmet for My Pillow*, wrote that he "could hear the enemy everywhere about me, whispering to each other and calling my name. … Everything and all the world became my enemy, and soon my very body betrayed me and became my foe" (Leckie, 2010, 73). Lt. Meehan's hearing experience was as dramatic, and the way he described it was as powerful. After a firefight, Meehan heard:

> … the sound of men in desperate pain. They weren't screaming. They weren't yelling or crying. They were making deep, guttural sounds that rose up from their lungs and their souls. The sounds weren't human. They were animalistic, primal, and desperate. They were loud, cutting through other sounds, rising up above the shouted orders and the crumbling house and demanding to be heard. I

hadn't heard anything like this. I hadn't seen anything like this and I was scared. (Meehan, 2009, 120)

These passages offered by veterans of several different wars tell the story of extremely demanding, stressful, and abnormal conditions. The body must adapt or deal with an environment that upsets and troubles nearly all senses at all times. Even the most realistic training programs fail to re-create conditions that fully resemble the demand of the battlefield on the individual. If for no other reason, soldiers put through demanding training are aware that the program will terminate in a matter of weeks. Former Marine Corps Captain Boudreau provides a strong explanation about the limitations of training. He stated that the training delivers individual soldiers and Marines "to the threshold of war, through that first shot down range, through that first kill. From that point forward, he finds himself moving into uncharted territory. It is up to him as an individual to negotiate all those moral obstacles on his own. The training means nothing anymore." (Boudreau, 2008, 82) In a senses-saturated battlefield, normalcy will not be restored until the end of the deployment, and that might be months away.

The Evil Zone: The Individual

The physical environment of the battlefield, either in a conventional war or a counterinsurgency campaign, is often abnormal and frightening. The absence of normal or even quasi normal living conditions is only part of the making of the physical environment. It is in such an environment that the greatest threats to soldiers' lives hide. Soldiers on the front line fear for their lives and for those of their friends and comrades; in a counterinsurgency operation, that fear is constant. While frontline soldiers might be able to step back from the reality of war when they are in the rear lines, soldiers operating in a counterinsurgency environment often will not enjoy such an opportunity. In addition, while the frontline soldier can easily identify the source of the threat he faces—an enemy with a uniform—in a counterinsurgency environment, anyone can be the enemy at any time. There is no pause from the fear of being killed.

Fear is indeed a key factor that has a major effect on individual soldiers' psyche and behavior. According to Brigadier General H.R. McMaster "[u]nits that experience the confusion and intensity of battle for the first time in actual combat are susceptible to fear. Fear can cause inaction or, in a counter-insurgency environment, might lead to an overreaction that harms innocents and undermines the counter-insurgent's mission" (McMaster, 2009, 16).

Bill Heflin, a Vietnam veteran who served as a "tunnel rat," arguably one of the most challenging tasks, provided a graphic description of fear. He said that "fear … can shake hips so bad that you cannot function" (*First Kill*, 2002). Caputo explains fear as "almost like a kind of chronic malaria. There are times when you overcome it completely, there are other times it will just wash over you and can almost feel paralyzed" (Inside the Vietnam War, 2008).

One of the greatest sources of fear is the unknown. In Vietnam, booby traps were infamous among soldiers as much as IEDs and VBIEDs are infamous among troops deployed in Iraq and Afghanistan. Indeed, the effectiveness of the more modern IEDs is significantly higher than the booby traps, but the fear among soldiers is arguably not much different. In Vietnam, the wrong step might have killed or badly mutilated soldiers; in Iraq, the wrong turn in a vehicle on a busy street might kill them. The experience of operating under such a pressure is psychologically exhausting.

For soldiers, such a physical environment represents an emotional and psychological challenge. Several authors have explained the emotional implications of operating under this constant fear. Jim Frederick in *Black Hearts* has provided an excellent and gripping description of such emotions. Frederick stressed how terrifying the experience of leaving the base can be, especially when soldiers are expected to spend the entire time out on patrol waiting for the explosion that might kill them. Every patrol is a gamble. Soldiers ride their Humvees with their "butt cheeks and fist clenched, doing deep breathing to get control of [their] heart rate and ... nausea the whole time, waiting for it." And when it does happen, it is just one instant.

> The power of the charge and where it hits has much to do with the aftermath. 'Sometimes you remember every millisecond of the thwomp. ....Other times you black out for those crucial few seconds.....what happened? How long did it happen? Am I okay? Is that other guy okay? Are we all okay?'

Frederick also noted that in the moments following the blast,

> the anger builds as you review what just happened. Somebody not far from this spot, someone right around here—it could be him, or him, or him, just tried to kill you. Who of these motherfuckers just tried to kill you? The people around the scene of the explosion, those civilians who have become accustomed to such terrible events are silent witnesses. Somebody tried to kill you, and all of these people know something, yet they are not saying anything. How could you not want to kill them too for protecting the person who just tried to kill you? How would you contain the rage? (Frederick, 2010, 84–5)

In such an environment, soldiers suspect everyone. According to Caputo "you start to suspect everybody, everybody, even children, and often not without cause do you suspect them. And suspicion is not that big a step from outright hatred." (*Inside the Vietnam War*, 2008). General Peers noted that, as a result of the booby traps and mines that caused several casualties among the men of Charlie Company in the months before the My Lai massacre, "[m]any of the men thought these devices had been laid by the women, children and old men or that if Vietnamese civilians had not planted them they at least knew where they were, but never warns the American troops." (Peers, 1979, 231)

Thus, in a COIN environment it is possible that the civilian population becomes the target of soldiers' rage. Lt. Meehan explained that behind rage and anger there is something significantly deeper: it is desperation.

> They [the soldiers in his unit] were desperate to avenge the deaths of their men, and they were desperate to survive. They were desperate to get home alive, and if one more dead Iraqi was the cost of that chance to get home, so be it ... I did not believe that we could all get to that point, but I feared it. (Meehan, 2009, 105)

Meehan described his rage following the explosion of an IED. "For the first time in my deployment to Iraq, indeed for the first time in my life, I felt rage boil inside me. That final IED brought out of me." He expressed his hatred for those who tried to kill him and his soldiers, and his uneasiness about being in Iraq. He continued:

> I had never felt that way. I did not know I was capable of it, and I did not know if in the coming days I would be able to lose that feeling and maintain the composure that I prided myself on ... All I knew, and all I felt, was the desire to inflict harm on the people around me. All I felt was a desire to protect myself and my men, and, as our tank bounced across the province on its way back to Warhorse, all I could think was that killing seemed like the right way to do it. (Meehan, 2009, 129–30)

Donovan Campbell, a small unit leader and a Marine Corps veteran of Iraq and Afghanistan described the impact fighting and taking casualties had on his Marines. In one incident, the Marines intervened to help a group of children who were badly hurt by an IED explosion. While the Marines were helping several injured children, they were attacked by a group of insurgents and Lance Corporal Todd Bolding was badly wounded. Bolding died later that day. At the end of the firefight, the Marines returned to base. When Lt. Campbell looked at his Marines, some were stunned and extremely somber. A few were talking quietly, others, however:

> ...stood by themselves with hard eyes and stone faces, fingering their weapons. Over time, I watched as more of my Marines joined the latter category ... They wanted revenge on our faceless enemies and on the fearful civilians whose hesitance had prolonged our waiting and cost us one of our best men. They wanted revenge on the stupid, broken Iraqi public services ... they wanted revenge on the miserable city of Ramadi for forcing us to make horrible choices, day in and day out, until it seemed like no matter what path we took, we lost. (Campbell, 2009, 229)

Lt. Campbell had in front of him a group of individuals who had been tested to the extreme and were now just one step away from entering the evil zone. In many cases, very likely in all cases, at this point it is the leader's wisdom and lucidity of mind that can prevent that last, final, crucial step from being taken. A few words can determine the descent of the group into the evil zone or can keep them away from it. One should bear in mind that even the best leaders need to constantly remain morally alert to the pressure of situational forces. They are the last and the most powerful protection in preventing soldiers and Marines from entering the evil zone.

Campbell shared the emotion and the rage, yet, he wrote "I was a lieutenant, and a leader, and no matter what I felt, I had to take care of my men and accomplish our mission, and, unfortunately, revenge wasn't our mission" (Campbell, 2009, 229). Campbell understood how difficult the moment was for his unit and how important it was for him to provide proper leadership. He addressed the Marines:

> Here's what we're not going to do. We're not going to kill everyone we feel like. We're not going to shoot indiscriminately at random civilians every time the fire breaks out. We've worked too hard to quit on the mission now ... And you know what? ... Bolding [the fallen Marine] wouldn't want us to start killing everyone randomly. You know Bolding, you know this is true. If he were here, you know that he would tell us never mind him, to keep doing what we were doing because it's the right thing to do. (Campbell, 2009, 230)

Indeed, there are countless occasions in which small unit leaders deployed in this environment do and say the right thing. They have the ability to control the potential for the group to develop a dynamic that might lead them into the "evil zone." They provide clarity and re-enforce the right course of action. The confusion that follows traumatic events like the death of Lance Corporal Bolding can generate extremely dangerous group dynamics.

The ambiguity of the situation raises key questions among the group that went through such a traumatic experience: Is it acceptable and fair that a Marine was killed while he was helping wounded children? Indeed, in some occasions there might be a high price to pay for doing what is right under abnormal circumstances. Yet, these are the circumstances that create great confusion and ambiguity. Steven Baum rightly stressed that "When reality becomes ambiguous and we become uncertain of our own judgments, we look to others for directions" (Baum, 2008, 86). It is at this point that leaders' clarity of mind and moral awareness are fundamental to preventing the group from stepping into the "evil zone."

The stressors and traumas of operating in a combat zone indeed have the potential to seriously compromise the individual's ability to retain his or her sanity. The most compelling evidence of such a situation is the growing number of veterans who suffer from PTSD and the increasing number of those who have committed suicide.

For nearly a decade, the U.S. Army MHAT (Mental Health Advisory Team) has conducted studies on the mental health and behavior of soldiers and Marines deployed in Iraq and Afghanistan. The report released in 2006—MHAT IV— included a chapter, "Battlefield Ethics," which provides survey responses from both soldiers and Marines who have served in a combat environment. The report is eye-opening. For the objective of this chapter, a few enlightening passages that show evidence of the connections between mental health, ethical behavior, and leadership should be noted.

According to MHAT IV "Soldiers who screened positive for a mental health problem (anxiety, depression or acute stress) were twice as likely to engage in unethical behavior compared to those Soldiers who did not screen positive." The likelihood of unethical behavior was more likely the outcome of mental conditions that psychologists might consider not as critical as PTSD. MHAT IV also noted that "Soldiers and Marines who had high levels of anger were twice as likely to engage in unethical behaviors on the battlefield compared to those Soldiers and Marines who had low levels of anger." It is very likely that individuals involved in combat will experience rage and anger. Therefore, the issue is not how to avoid rage, because at some point and some time they will experience such emotions; the issue is how to control anger and rage. Individuals who operate in such a situation should learn how to avoid becoming blinded by such emotions.

In addition, MHAT IV emphasized that for soldiers and Marines "having a unit member become a casualty or handling dead bodies and human remains were associated with increases in the mistreatments of Iraq (sic) non-combatants." Indeed, it is more likely after the killing of a comrade that a small unit might cross into the evil zone. The loss of a brother in arm is a major trauma for a unit that often is already experiencing high level of stress. In many cases, the killing of a close friend happens under dramatic circumstances. As Sledgehammer and Caputo explained, death under those circumstances is extremely powerful, indeed a major trauma. Lt. Campbell managed the death of Bolding and its consequences on the unit in an outstanding way. It is likely that hundreds of small unit leaders have been able to deal with such a difficult situation as well as Lt. Campbell did. Indeed, one of the most important outcomes provided by MHAT IV was that it highlighted the importance of small unit leadership. MHAT IV noted that "Soldiers who reported that they had good junior NCO leadership reported higher morale and fewer mental health concerns. Good junior officer leadership was also associated with Soldiers and Marines following the Rules of Engagement (ROE)."

"Situational Awareness" and the Evil Zone

The "evil zone" is a space in which any individual, and particularly those individuals who have stressful professions, such as military personnel, might step into. There are individuals who are more resilient than others, but there are probably few who are totally immune from the possibility of entering the evil zone.

In particular, those in a leadership position should always be open to the possibility that moral mistakes might be made, that they operate in spaces that border with the evil zone, and that immunity from perpetrating evil is a dangerous belief that might create a sense of false security. Zimbardo rightly stressed that "Paradoxically, by creating this myth of our invulnerability to situational forces, we set ourselves up for a fall by not being sufficiently vigilant to situational forces" (Zimbardo, 2008, 211).

Philip Caputo provided an elegant and effective description of the power of the situation in Vietnam. The Marine veteran wrote that in such an environment "Everything rotted and corroded quickly over there: bodies, boot leather, canvas, metal, morals." It is enlightening that an individual as attentive and sensitive as Caputo places morals among the many "things" that were corroded by the environment. Caputo continued "Scorched by the sun, wracked by the wind and rain of the monsoon, fighting in alien swamps and jungles, our humanity rubbed off of us as the protective bluing rubbed off the barrels of our rifles" (Caputo, 1996, 229).

Therefore, the purpose of exploring and deconstructing the evil zone is to identify a "space" in which any individual, even a strong individual of character might step into and might act in an unexpected, abnormal, and possibly evil way. The study of the evil zone should provide those individuals who might be acting in such a space with a strong mind set. They should develop a broader and more comprehensive understanding of what "situational awareness" implies, and be aware of the power of "situational" forces and the impact they might have on their behavior. There is no doubt that those forces can be properly dealt with, yet individuals will be better prepared to deal with them if they acknowledge and understand their power. A key question is whether individuals might be more prone to enter the evil zone if they delay the response to a certain situation or if they act immediately. This is an issue that needs much research, but here I will try to explain why I am more inclined to believe that an immediate reaction might be ethically more effective than a delayed one. In this context, immediate does not mean instinctual. An immediate reaction should be the outcome of a rational thinking process, when indeed emotions are at play, but they have not yet compromised the individual's ability to identify the right course of action. For instance, if individual soldiers know that the right thing to do is not to torture a POW, they will probably refrain from engaging with the idea of harming the POW. However, what if a fellow soldier has been apprehended by an insurgent group to which the POW belongs? What if that soldier is the most liked by the members of the unit? What if they begin to believe that by using a "little" violence on the POW they might gain some good information that will help them to free the fellow soldier? And what is their state of mind if just a few days before this event the bodies of two fellow soldiers had been found with evident signs that they had been tortured before being killed by the enemy? How many other considerations and events can make an emotional impact and affect the ability of soldiers to do what is right—not to torture the POW?

It might be dangerously easy to step into the "evil zone" whenever the normal moral decision making process is seriously delayed and could become increasingly more confused. The confusion might affect the individual's ability to understand and separate right from wrong. Bill Heflin said that while he was in Vietnam "It was like there was the right and there was the wrong, but it did not seem that the wrong was that far away from the right … and it did not feel no different" (*First Kill*, 2002).

Indeed, Caputo and Heflin description of their experience in Vietnam provides us with an interesting reflection on a significant aspect of the famous experiment on obedience conducted by Stanley Milgram in the 1960s (Milgram, 2004). In the experiment in which a teacher (the subject of the experiment whose behavior was monitored by Milgram and his research team) was progressively ordered to inflict what he believed to be an electric shock to a learner. The shocks were inflicted gradually. Initially, the teacher inflicted what he believed to be an extremely mild shock. We are left wondering whether the same number of individuals (65 percent) would have inflicted the strongest, most painful and maybe deadly, shock immediately. The teacher was gradually pushed from what looked like an unobjectionable behavior—inflicting a mild punishment—to causing severe pain that bordered on killing (Flanagan, 1991, 298–300). This idea of graduality is clearly identifiable in the "progression" towards the evil zone. A certain behavior that might be unacceptable at the beginning of the deployment, a few months into the mission might be mildly punished, and later on might be tolerated and or accepted. In the case of the My Lai massacre, it was over the course of several months that the troops began to gradually brutalize civilians. The leadership, which is the most effective in stopping such a progression, accepted and then encouraged such a behavior. Greg Olsen a machine gunner with Charlie Company in a letter to his father from Vietnam wrote:

> One of our platoons went on a routine patrol today and came across a 155-millimeter round that was booby-trapped. Killed one man, blew the legs off two others, and injured two more. On their way back to the LZ, they saw a woman working in the fields. They shot and wounded her. Then, they kicked her to death and emptied their magazines into her head. It was murder; I'm ashamed of myself for not trying to do something about it. This isn't the first time, Dad. I've seen it many times before. My faith in my fellow man is all shot to hell. (My Lai, 2010)

At the end of the My Lai investigation, General Peers identified the factors that, in his view, might have contributed to the massacre. In relation to the idea of graduality, it should be noted that General Peers stressed that "well before My Lai operation there had been instances of mistreatment, rape and some unnecessary killings in Task Force Barker" (Peers, 1979, 232).

Difficult, stressful, demanding, and fundamentally abnormal situations, can delay individuals' decision making process, even though they know what the right course of action is. It is in this delay that the possibility of entering the evil zone increases. That delay is indeed generated by the impact of emotional, sociological, and psychological forces that are particularly strong in an abnormal environment.

Therefore, there are courses of action that should be treated as absolute. For these courses of action, training programs that provide immediate responses very likely are the most effective. However, individuals who will operate in stressful, abnormal situations should be educated about the power of situational forces and develop a strong understanding of who they are and how they operate under such a pressure. Ultimately, when confused they will look to their leaders to find the answers they need.

In many ways, the role of leaders in an ambiguous situation is pivotal because they are supposed to provide additional guidance on what the right course of action is. When deployed in a COIN environment, it is not unusual that individuals might struggle to find an "identifiable and reliable" point of reference that will guide their behavior. In a COIN "battlefield" those points of reference provided by society might be absent or significantly compromised. Herman Langner a psychiatrist who treated Vietnam veterans, in relation to "Bob" a Navy corpsman who had been involved in an "incident" similar to My Lai wrote: "Under ordinary circumstances he [Bob] would have expressed his aggressions and insecurities as most of the rest of us do in our daily dealings with our fellow man. The combination of these aggressions and insecurities and the pathological circumstances in Viet Nam made him a murderer" (Langner, 1971, 953). "Bob," like many other young individuals deployed in a COIN environment, having no societal point of reference to guide his moral behavior, looked at his peers and leaders; clearly the point of reference they provided him was corrupt.

Bibliography

Baum, S. (2008), *The Psychology of Genocide* (Cambridge: Cambridge University Press).

Boudreau, T. (2008), *Packing Inferno: The Unmaking of a Marine* (Port Townsend, WA: Feral House).

Britt, T.W. and Adler, A.B. (2003), *The Psychology of the Peacekeeper: Lessons from the Field* (Westport, CT: Greenwood).

Browning, C.R. (1993), *Ordinary Men: Reserve Police Battalion 101 and the Final Solution in Poland* (New York: Harper Perennial).

Campbell, D. (2009), *Joker One: A Marine Platoon's Story of Courage, Leadership, and Brotherhood* (New York: Random House).

Caputo, P. (1996), *A Rumor of War* (New York: Owl Books).

Card, C. (2002), *The Atrocity Paradigm: A Theory of Evil* (Oxford: Oxford University Press).

Finkel, D. (2009), *The Good Soldiers* (New York: Sarah Crichton Books).

First Kill (2002), Directed by Coco Schrijber, First Run/Icarus Films.

Flanagan, O. (1991), *Varieties of Moral Personalities* (Cambridge, MA: Harvard University Press).

Frederick, J. (2010), *Black Hearts* (New York: Harmony Books).

National Geographic (2008), *Inside the Vietnam War.*

Hinton, A.L. (2005), *Why Did They Kill?: Cambodia in the Shadow of Genocide* (Berkeley and Los Angeles: University of California Press).

Langholtz, H.J. (1998), *The Psychology of Peacekeeping* (Westport, CT: Greenwood).

Langner, H.P. (1971), "The Making of a Murderer," *American Journal of Psychiatry*, 127:7, 950–3.

Leckie, R. (2010), *Helmet for My Pillow* (New York: Bantam Books).

Lifton, R.J. (1998), "Bearing Witness to My Lai and Vietnam," in Anderson, D. (ed.) *Facing My Lai: Moving beyond the Massacre* (Lawrence: University Press of Kansas).

McMaster, H.R. (2009), "Preserving Soldiers' Moral Character in Counter-Insurgency Operations," in Carrick, D., Connelly, J. and Robinson, P. (eds), *Ethics Education for Irregular Warfare* (Farnham: Ashgate).

Meehan, S. and Thompson, R. (2009), *Beyond Duty: Life on the Frontline in Iraq* (Cambridge: Polity Press).

MHAT IV Report available at the US Army Medical Department, Army Medicine, at http://www.armymedicine.army.mil/reports/mhat/mhat.html.

Milgram, S. (2004), *Obedience to Authority: An Experimental View* (New York: Perennial Classics).

My Lai (2010), PBS, American Experience, Transcript of the documentary available at http://www.pbs.org/wgbh/americanexperience/features/transcript/mylai-transcript/

Newman, L.S. and Erber, R. (2002),*Understanding Genocide: The Social Psychology of the Holocaust* (Oxford: Oxford University Press).

Osiel, M. (2009), *Making Sense of Mass Atrocity* (Cambridge: Cambridge University Press).

Peers, W.R. (1979), *The My Lai Inquiry* (New York: Norton & Company).

Remember My Lai (1989), Frontline, PBS. Transcripts available at http://www.pbs.org/wgbh/pages/frontline/programs/transcripts/714.html.

Shay, J. (1995), *Achilles in Vietnam: Combat Trauma and the Undoing of Character* (New York: Simon & Schuster; First Touchstone edition).

Sledge, E.B. (2010), *With the Old Breed* (New York: Presidio Press).

Tripodi P. (2010), "Understanding Atrocities: What Commanders Can do to Prevent Them," in Whetham, D. (ed.), *Ethics, Law and Military Operations* (Basingstoke: Palgrave).

Tripp, E.F. (2008), *Surviving Iraq: Soldiers' Stories* (Northampton: Olive Branch Press).

Waller, J. (2007), *Becoming Evil: How Ordinary People Commit Genocide and Mass Killing* (Oxford: Oxford University Press).

Zimbardo, P.G. (2008), *The Lucifer Effect: Understanding How Good People Turn Evil* (New York: Random House).

Chapter 12

Psychological Foundations of Unethical Actions in Military Operations

J. Peter Bradley

As the title of this book suggests, the chapters in this volume explore the changing nature of war and the implications of "new wars" for military ethics. One aspect of war that has evolved in recent times is an increased emphasis on the actions of the individual soldier.[1] Part of this trend is due to greater public interest in human rights and concern for the safety of noncombatants in war zones. Another part derives from the enhanced lethality of weapons systems and the shift to smaller armies. Almost all of the militaries in the western world have moved from conscript armies to professional forces with the result that military forces are now smaller than they once were and soldiers are more valuable, because they are fewer in number and cost more to arm and train. Military tactics have also changed with smaller formations now dispersed more widely on the modern battlefield so that soldiers and junior leaders are now operating with greater levels of autonomy than in earlier conflicts (Krulak, 1999).

The ascendancy of the individual soldier is not without risks. Fewer soldiers means that more is expected of each one and we have already witnessed the disproportionate effect a few individuals can have when they fail to live up to ethical and professional military standards. The abuse of Iraqi detainees at Abu Ghraib (Danner, 2004; Zimbardo, 2007) was influenced, at least in part, by interrogators, described by some in the Pentagon as "the six guys who lost us the war" (Streatfield, 2007, 378). These were the individuals who used torture and encouraged guards to rough up the detainees in preparation for interrogation (Danner, 2004, 8). In a lesser-known incident, a handful of paratroopers in the Canadian Airborne Regiment on operations in Somalia in 1993 killed several local men under dubious circumstances and beat another to death (Canada, DND, 1997; Bercusson, 1996). The Canadian Government reviewed the harmful actions of its paratroopers in Somalia along with other incidents of unprofessional behavior in the same unit and then disbanded the tainted Canadian Airborne Regiment, to the disappointment of many within and outside the military community.

Professional and ethical failures, such as those mentioned above, typically begin with a few individuals. In fact, we are often quick to blame such events

---

1   By soldier I mean any member of a military force serving in army, navy or air force units.

on "the few bad apples," but these incidents cannot occur without the complicity of others, either passive complicity in the case of bystanders who observe but fail to intervene, or active complicity by observers who actually participate in the harmful actions. When such events occur, many of us are mystified as to what went wrong, but there is actually a substantial body of scientific theory and research that can explain much of this behavior. Following this line of inquiry, the aim of this chapter is to present a behavioral science perspective on moral decision making to show how moral failings can occur in military operations and to examine certain aspects of the military environment that can actually contribute to unethical behavior.

The thrust of this argument goes like this: On today's battlefields military forces need soldiers who can make sound moral decisions.[2] Unfortunately, humans are naturally susceptible to decision-making errors and biases. Moral competence is properly viewed as a multidimensional concept involving a mixture of cognition, motivation and emotion which further complicates moral decision making. Individual soldiers are responsible for their actions, but because of intrapersonal and interpersonal differences, some will be stronger than others in some areas. The major message in this chapter is that situational factors like organizational culture and small-group processes exert enormous influence on the choices individuals make and such elements are particularly powerful in military settings, especially in combat. Many of the team-building processes that are effective and actually essential for success in battle can have a deleterious effect on the moral competency of soldiers. Certain elements of military culture interact with basic psychological processes which, if not properly managed, can lead to unethical behavior. At the centre of this complex mix of influences and potentially harmful outcomes is the individual soldier, of whom military leaders and the nation's citizenry expect so much. Unfortunately, many young men and women in uniform may not be adequately prepared to navigate their way through these ethical challenges.

Decision Making

According to psychologists who study in this area, individuals generally employ one of two approaches to decision making: a rapid, largely intuitive process, and a slower, more contemplative approach. The intuitive approach is the automatic, reflexive process we employ when making split-second decisions, as when we see a flashy automobile or an attractive person for the first time. In such cases, we make the decision automatically, on the basis of gut-feel, virtually without thinking at all. At other times we take a more deliberate, rational approach that is better suited for considering important factors and evaluating different options as we search for the best solution.

---

2   The terms ethics and morals are used interchangeably in this chapter (e.g., an ethical decision is a moral decision, moral issues re. ethical issues.

Some researchers believe that the reflexive process dates back to our earliest evolutionary roots, while the deliberate process is one which we humans developed later in our biological timeline (Hauser, 2006, 122; Marcus, 2008). As for which one we use the most, Jonathan Haidt, a researcher in this field contends that the, "ancestral (i.e., reflexive) system seems to be the default option, our first recourse just about all the time" (Haidt, 2001, 815).

Although the reflexive system is widely used by individuals, it is not known whether it is more accurate than the deliberate system. In his treatment of this topic, Gladwell (2005) gives examples of rapid decision making successes and failures. Reflexive decision making seems most effective in routine or familiar situations, and deliberate decision making is thought to be more effective in novel situations where we need to devote some time to assess the problem before us. Interestingly, this mix of reflexive and deliberate decision making is recognized in the way planning is taught in military training courses. Instructors emphasize rational decision-making conventions like the combat estimate and operational planning processes in military courses so that students will over-learn these procedures to the point that they become automatic responses when needed later in operations.

Although recent research has shown that automatic decision making is more common and more effective than previously thought (Damasio, 1994), there is still a great deal of attachment to rational decision making and a reluctance to advocate only intuitive decision making, particularly in the case of moral decisions. Intuitive decision making seems irresponsible when the lives and property of people are at risk.

A major problem with decision making is that although people make many decisions in their daily lives, the mixture of emotion, reason, biological drives and social influences that are involved can lead to errors in judgment. In fact there are so many types of possible errors they cannot be covered adequately here, but I will briefly summarize the field (Merkhoffer, n.d.). First, people are "cognitively lazy" by nature; they prefer to stay with what they already know and are reluctant to step outside their comfort zone to consider new information or new ways of solving problems. Second, people are susceptible to perceptual distortions, focusing on some things and excluding others. One of the big perceptual distortions we frequently fall victim to is self-justification to soothe our ego after making a contentious decision (Tarvis and Aronson, 2007, 6). Third, the judgment of people can be distorted by their motivations or by incentives in the environment. Fourth, people can engage in faulty reasoning. Fifth, when working in groups, individuals can have their perceptions, motivations and reasoning distorted by influences emanating from the group. One of the more ironic findings from decision-making research is that many of us recognize that others are prone to biases, but believe we are not.

The Moral Domain

Moral functioning is about more than observable behavior. Actions like observing, thinking, and analysing are essential to moral functioning, but they are largely invisible. A useful framework for illustrating the breadth and complexity of moral functioning is the four-component model of James Rest (1986) and colleagues (Narvaez and Rest, 1995). The model consists of four elements: (1) Moral sensitivity, sometimes called moral awareness or recognition, refers to the ability to perceive the moral dimensions of a given situation. (2) Moral judgment refers to the mental processes people use in evaluating the choices open to them. Because this component draws heavily on one's cognitive abilities, more intelligent individuals tend to make better moral choices on average (Kohlberg, 1976, 32). (3) Moral motivation refers to the inducements people draw on when taking moral, or immoral actions. People who are spurred by moral or professional considerations are said to be morally motivated, whereas those acting on self-serving goals are not. (4) Moral action/character is the most complex of the model components as it blends personality attributes and behaviors.

Each of these components can be viewed in three ways: (1) as a single part of the broad domain of moral functioning, (2) as one step in a four-step process of moral functioning, or (3) as a personal quality that contributes to moral actions (Walker, 2004). The implication of viewing these components as personal qualities is useful in that deficiencies in any component might be developed through training or education. That said, research has shown that moral sensitivity and judgment can be developed through education (Bebeau, 2002; Colby, 2008; Pascarella and Terenzini, 1991, 2005), but there is little empirical evidence that motivation and action can be enhanced the same way.

Moral dilemmas are problems in which a person must choose between competing moral obligations or principles. A typical moral dilemma presents an individual with two or more moral obligations, but taking action to satisfy one moral obligation necessarily means neglecting the other morally-obliged action or actions (Stanford Encyclopedia). There are many moral dilemmas in military life.

A classic moral dilemma for military leaders is the conflict between troop safety and mission success. All military leaders are obliged to obey lawful orders to the best of their ability. This means that they should lead their troops aggressively in the pursuit of military objectives, assuming of course that the mission is a lawful one (most military forces have laws forbidding soldiers from following illegal orders [Osiel, 2002, 41]). But soldiers can be injured or killed while trying to accomplish their mission and all leaders have the moral obligation to preserve the safety of their subordinates. Some might say that this is not really a moral dilemma, that mission success always trumps troop safety when the two are in conflict, but this does not seem to be the case in modern operations, particularly humanitarian missions (e.g., peacekeeping) in which national interests are not readily apparent to the soldiers and junior leaders whose lives are at immediate risk. In fact, a Canadian study, conducted in the 1990s, showed that 30 percent of 600 military leaders who had served abroad on operations reported disobeying orders or moderating their response to orders to ensure troop safety (Canada, DND, 2001, 11).

Determinants of Moral Behavior

It is widely accepted by social scientists that the behavior of individuals is largely influenced by a combination of personal characteristics and situational factors. At the same time, these forces are so intricately intertwined it is often difficult to determine precisely the relative impact of the individual and the situation. Let us now begin by examining the role that characteristics of the individual play in ethical functioning.

Individual Factors

For many people life is a constant struggle between altruism and self-interest. We want to help others, but at the same time, we need to take care of ourselves and loved ones. When we find it difficult to be altruistic, many of us give in to environmental pressures and take the easier path. But some individuals are able to surmount situational influences. Take Private Harry Stanley, for example, and other soldiers in Lieutenant Calley's U.S. Army platoon who had the moral courage to refuse orders to shoot noncombatant women and children in My Lai, South Vietnam, on that murderous day in March of 1968. How was Stanley able to resist the orders to kill while other soldiers in his platoon were either unable or unwilling to do likewise? It is difficult to know for sure. It may have been values, but as Stanley himself said of his platoon mates, "I know that they had to have the same values as I had somewhere along the line" (Bilton and Sim, 1992, 373). It is clear that Stanley knew the orders should not be followed, "... ordering me to shoot down innocent people, that's not an order—that's craziness to me, you know. And so I don't feel like I have to obey that" (Bilton and Sim, 1992, 19). If Haidt's (2001) theory, mentioned previously, is correct, Stanley may not have thought about the order very much at all. He may have determined instantaneously that the order was wrong and should not be obeyed.

Scholars who study ethical decision making have identified a number of personal characteristics that are associated with taking ethical action. The most important are the moral competencies mentioned earlier (moral sensitivity, judgment, motivation, and character) and the personality traits of empathy, locus of control, responsibility and self-efficacy (Dovidio, Piliavin, Schroeder and Penner, 2006). It should be noted that, for the most part, this research has been conducted with civilian subjects, so it is unclear to what extent these findings generalize to the military milieu.

Moral competence   Of the four moral competencies mentioned earlier, the most conclusive research findings relate to moral judgment, showing that higher levels of moral judgment lead to greater probability of moral behavior (Thoma, 2002, 241). There are a number of procedures for measuring moral judgment, most of them based on the theoretical framework of Kohlberg (1976), a pioneer in this field. According to Kohlberg, those at the lower end of the moral judgment scale determine what is right or wrong based on the potential consequences to themselves (Will I benefit or be harmed as a result of my action?). Individuals at the next level determine right or wrong based on their desire to conform to group norms and rules (What will my peers think? What do the regulations require?). Individuals at the highest level of ethical reasoning use higher-order principles (e.g., justice, human rights) to determine what is right and wrong. Most adults reside in the middle group, where they determine what is right and wrong according to group norms and regulations. Given the emphasis on maintaining group norms in the military, it is not entirely clear if ethical decision making based on higher-order principles could flourish in military units. In fact, studies show that the average scores of military samples on tests of moral reasoning ability typically fall in the middle range, reflecting a "maintaining norms" approach to moral reasoning (Rest, 1994, 14).

Empathy   Ethics involves taking the interests of others into account, so it is not surprising that researchers find that empathy is related to ethical functioning (Rest, 1994). Based on these results, we would expect soldiers with empathy to display higher levels of moral competence. For example, empathic soldiers should be able to perceive the plight of others (moral sensitivity), consider the relevant issues more deeply, focus more on likely consequences to stakeholders, and make sound moral choices (moral judgment). Greater empathy should also lead soldiers to be more interested in relieving the distress of others (moral motivation). There may be a potential obstacle to soldier empathy, however, in the socialization processes employed to develop group cohesion in the military, for unit cohesion is often achieved at the expense of reduced consideration for out-group members.

Locus of control   Individuals with an internal locus of control believe that their behavior has an impact, that what they do matters. Research has shown that individuals possessing an internal locus of control are more likely to engage in ethical behavior (Forte, 2005). Individuals with an external locus of control believe that their actions have little effect on outcomes, so they are more susceptible to external factors and more likely to succumb to unethical influences.

Responsibility   Responsibility is related to ethical behavior in that responsible individuals are more inclined to take action when things are not right. The same is true for individuals possessing high levels of self-efficacy, a related quality. Self-efficacious people are confident and action-oriented. Accordingly, people who are responsible and self-efficacious are more likely to take ethical action than individuals with low self-esteem who are more likely to become passive followers in ethically-charged situations (Aronson, 2004, 21).

Moral agency is related to responsibility. People who have moral agency can analyse a situation and then take action that is consistent with their moral judgment and motivation. By contrast, individuals who defer to the moral authority of others can slide into what social scientists call an agentic state where they follow the moral demands of another. It is easy to dismiss this as an unlikely scenario, but it is the way that many unethical incidents occur. The psychology of this process was dramatically captured in the experiments of Stanley Milgram (2004) in which subjects working under the direction of a university professor gave (what they thought were) electric shocks to another person in an experiment on learning. Although they became increasingly anxious and distraught with their actions, especially when the person receiving the shocks (who was an actor playing the role of learner, no shocks were actually given) started to complain, two-thirds of the subjects administered shocks long after the victim stopped screaming. When asked later why they did it, they reported that the experiment and the professor demanded it. They just followed orders. Where have we heard this before? Essentially, the subjects had shifted their moral attention away from the implications of shocking another human being to a moral consideration of how well they were living up to the expectations of the authority figure (i.e., the professor) in the experiment (Milgram, 2004, 7 8). Given the explicit lines of authority in the military, moral agency and deference are profoundly important in military ethics, and will be examined further in the final part of the chapter.

Situational Influences

Environmental influences on the behavior of individuals are extensive and more powerful than most of us realize. They range from macro-level factors like cultural and ideological influences to organizational (institutional) characteristics to attributes of the immediate situation and the social processes of primary groups, those small military groups "characterized by intimate face-to-face association and cooperation" (Manning, 1991, 457).

Macro-societal factors   There are many historical examples of societal-cultural-ideological influences leading to unethical outcomes. Examples include the Nazi holocaust (Kamenetsky, 1997), the Rwandan genocide (Hintjens, 1999), and the recent wave of Islamic fundamentalism (Ben-Dor and Pedahzur, 2004). A recent book by Martin Shaw (2005) on risk-transfer war illustrates how culture and ideology can influence the ethical behavior of soldiers in present-day operations. Shaw shows how, over the last few decades, democratic ideology and its inherent requirement for elected officials to maintain the support of their citizenry, has led western governments and their military forces to keep the true costs of war hidden from their populations. This is achieved in several ways, each of which can lead to unethical outcomes. First, keeping casualty rates of one's own forces low is achieved by transferring the risk (of death, injury, and destruction of property) to one's opponents, particularly the civilian population of the opposing side, effectively ignoring jus in bello obligations towards noncombatants. High-level bombing and use of overwhelming force (e.g., shock and awe) ensure that friendly ground troops will suffer fewer casualties, but these actions violate the jus in bello obligations of discrimination and proportionate use of force. Second, the media are managed so that journalists have limited access to information on casualties and destruction. Pooling journalists, as in the Gulf War, ensures that they are restricted in where they go and what they see, and embedding journalists with combat units as in the post-9/11 operations in Iraq and Afghanistan ensures that journalists see the war through the eyes of friendly soldiers and report favorably.

So, what are the ethical implications of risk-transfer war? First, the populations of western nations do not get an accurate view of the destruction their governments are inflicting on the other side and individuals working in the government and military bureaucracies must compromise their integrity to keep the true costs of war concealed from the public. Second, risk-transfer influences can impact unit commanders and soldiers who see that the lives of the enemy do not have the same value as the lives of friendly forces. Of course, soldiers have always seen their lives and the lives of their comrades as more important than the lives of enemy combatants and noncombatants, so they will take comfort when this sentiment is expressed by their military and political leaders. Ethically ambiguous or unethical acts can follow as soldiers and commanders strive to keep friendly force casualties low to satisfy both the wishes of the nation and its military personnel. For an example of this effect, Smith (2008, 154) reports how the desire to avoid military casualties led to increased civilian casualties during checkpoint operations in Iraq because soldiers interpreted rules of engagement to emphasize force protection over noncombatant safety. Similarly, the regular use of military aircraft to drop powerful bombs on small groups of enemy soldiers during post-9/11 operations in Afghanistan also reflects this desire to promote the security of friendly forces, even at the risk of endangering noncombatants.

Societal-cultural influences likely contributed to the unethical behavior of certain Canadian soldiers on operations in Somalia in 1993. Members of the Canadian Airborne Regiment arrived in Africa believing they would be helping Somali society recover from civil war, but when they experienced hostile reactions like rock throwing and spitting from the people they were there to help, the soldiers grew frustrated. In addition, unfamiliar Somali cultural practices, like men holding hands and talking among themselves while the women worked—at tasks that would be considered men's work in Canada—might have prejudiced the Canadians' attitudes towards Somali men and set the stage for the unethical actions observed in that deployment (Shorey, 2006, 199). Principal among these transgressions were the killing of four Somalis under questionable circumstances and the torture and beating-to-death of a Somali teenager. Later, after homemade videos became public showing soldiers of the Airborne Regiment engaging in degrading and racist behaviors, the Canadian Government disbanded the Regiment in January 1995.

The Canadian Airborne Regiment had been formed in 1968 from the parachute companies of infantry regiments in the Canadian Army. Things went well in the early years, but because it never had its own recruiting system, the Regiment had to rely on other army units to provide soldiers, NCOs and officers for tours of two-three years. In the 1980s there was a noticeable drop in quality of the personnel being sent to the Regiment and discipline problems followed. These problems were serious enough that the army convened several investigations in the 1980s and early 1990s (for more on this see Horn, 2001, 2007).

Organizational factors   Organizational psychologists and sociologists have long known that organizations have the potential to influence individuals to counterproductive ends (Vaughan, 1999). In particular, organizational characteristics like reward systems, organizational culture and work processes (Trevino, Weaver and Reynolds, 2006, 966), although designed to promote efficiency and productivity, can also lead to second- and third-order consequences that can be harmful to individuals and organizations alike. I will show in the following paragraphs how these factors contributed to the unethical behavior of Canadian paratroopers in Somalia in 1993.

The Canadian Airborne Regiment was considered an elite unit and aggressive soldiering was encouraged and rewarded. This sometimes meant that infractions were overlooked by leaders, particularly if the incidents were considered minor and not directly related to field soldiering, or were committed by someone considered to be a good soldier. At the same time, social cohesion was very strong and everyone—officers, noncommissioned officers (NCOs), and soldiers alike—felt pressure to fit in. As a result, the tendency to dismiss some shortcomings and the desire to be part of the unit contributed to a lackadaisical approach to military discipline in the Regiment on occasion.

The organizational culture of the Regiment was shaped by an emphasis on the physical demands of soldiering "in the field" and parachuting. Regimental norms emphasized intense physical activity, macho behavior, and a disdain for those aspects of military life associated with the garrison (e.g., matters like saluting, administration, maintenance of equipment). Unfortunately, many of these aspects of military life are important for organizational effectiveness and military discipline.

An unhealthy mix of poor role models and strong cohesion also led to problems. Ideally, role models in a military unit should be leaders within the chain of command or exemplary soldiers in the ranks, but a number of the role models in the Regiment were not exemplary soldiers at all. Some were troublemakers or mediocre soldiers who were not wanted back in their original units and so remained in the Regiment longer than normal. Given the emphasis on parachuting in the Canadian Airborne Regiment, many of the role models were soldiers and NCOs who had been in the Regiment the longest and therefore had the most parachuting experience. As a result, some poor-quality personnel became role models for young, impressionable soldiers who could not distinguish between aggressive soldiering and lack of discipline. Moreover, outside influences were shunned because of the strong cohesion in the Regiment, so there was little encouragement to reflect on the interests or perspectives of others. Instead, the ethical perspective of many junior soldiers was shaped by the macho norms advocated by junior leaders and peers who had more time in the Regiment. The Canadian public had a glimpse of this effect when a homemade video of an Airborne hazing ritual aired on Canadian television in 1995 showing drunken paratroopers in disgusting behavior under the supervision of senior soldiers and NCOs.

In military organizations the requirements of obedience are clearly outlined in regulations and orders, and reinforced by leaders in the chain of command. In the Canadian Airborne Regiment there was another line of obedience in which loyalty was directed inward towards the "Airborne ethos" and the informal authority of role models outside the formal chain of command. This situation was aggravated by the fact that the officers and most of the more effective NCOs were posted into the Regiment for only two or three years. Being new to the airborne environment, some of them may have been awed by the mystique of serving with elite soldiers, felt tentative in their leadership roles and adopted the airborne practice of overlooking minor lapses of discipline.

When soldiers realize that they will be held responsible for their actions, they are more likely to make correct, ethical choices. As previously mentioned, the lax approach to discipline in the Canadian Airborne Regiment meant that minor infractions were occasionally overlooked. Soldiers also learned that they could thwart the chain of command and avoid responsibility for some of their unprofessional behavior by closing ranks. In one incident an officer's car had been set on fire, yet no one was held accountable because all the witnesses refused to cooperate with the military police investigators (Winslow, 2004, 12).

Group processes   Much of military work is team-based because the efforts of many are usually more effective than those of a single person and soldiers find comfort in the presence of others when in danger. While high-performing groups are essential for military effectiveness, there is a dark side to group work, as people will do things in a group they would never do by themselves (that's what makes youth gangs so dangerous). Consequently, it is important for soldiers and small-unit leaders to be aware of the harmful influences that can emerge in group work. I will now highlight five social processes that are problematic for military groups—conformity, deindividuation, bystander effect, groupthink, and moral disengagement.

Conformity   People are continuously torn between satisfying their need to be an individual and their need to conform to social norms, and conformity often overrides individuality in social situations. Research has shown that conformity is more likely to occur in groups that contain experts, where there is a lot of cohesion within the group, where members are committed to the group, where the other members of the group are important to the individual or are similar to him or her, and conformity is strong if the group typically holds people accountable for their actions (Aronson, 2004, 20). The inclination to conform, particularly in cohesive and homogenous groups as in the military, can aggravate ethnocentric and xenophobic tendencies, making conformity potentially fertile ground for unethical behavior. The tendency to conform is so strong that some individuals will participate in group behavior that they know is wrong and will later reframe it as correct behavior to reduce the dissonance their actions have caused them. Canadians witnessed the power of conformity in the previously-mentioned videos depicting Canadian Airborne Regiment soldiers behaving badly. One of the videos showed a hazing ritual in which one of the initiates, a black soldier, was on all fours, shirtless with KKK (the Ku Klux Klan is a white-supremacist group in the United States) printed on his back, being led around on a leash by another soldier. Remarkably, the initiate later defended both the initiation and his participation in public.

Deindividuation Deindividuation is a de-personalizing process in which individuals lose sight of their identity in the group and which can also lead to antisocial and unethical behavior. The process begins with individuals becoming emotionally aroused (e.g., excited, angered, frustrated) from the effect of the group's social dynamics, and replacing their sense of self-identity with a feeling of anonymity. At the same time, their sense of responsibility is diffused among the group, and they feel less accountable for their actions and more likely to behave in ways they never would have if alone. In this way, the process of deindividuation is marked by diminished self-awareness and self-restraint, and increased responsiveness to the demands of the immediate situation (Zimbardo, 2007, 305).

Bystander effect The bystander effect refers to the failure of people to intervene when they see someone in distress and are not sure whether they should intervene or not. This likely played a role in the death of a Somali teenager by soldiers of the Canadian Airborne Regiment during its peacekeeping mission in Somalia in 1993. The youngster had been caught stealing supplies and was held in a bunker by several paratroopers who beat him to death over the course of several hours. As many as 17 soldiers and NCOs visited the bunker during this period; others overheard what was going on, but no one intervened, perhaps because they thought that unit leaders and duty personnel who were in the vicinity could, would or should intervene (Shorey, 2000–2001, 24). Situations like this one show that individuals look to others for cues on how to act. Especially in ambiguous situations, individuals will look for role models within the group. In fact, studies of the bystander effect have produced the most counterintuitive of findings: there is a greater chance of someone in distress receiving help from a lone bystander than a group of bystanders (Zimbardo, 2007, 315).

Groupthink   Groupthink occurs when members of a group place more emphasis on protecting the social stability of the group than on making effective group decisions (Janis, 1972). Essentially, the group loses sight of the primary aim of the group's activity, which is to take effective action, and focuses instead on group harmony (Aronson, 2004, 15). Dissent is discouraged and individuals appear to agree with others even when they do not. If no one takes charge, the group can make poor or unethical decisions. Because small military units typically have strong in-group relations, groupthink is a potential threat to ethical functioning, particularly when strong, ethical leadership is not present.

Moral disengagement   People are able to keep their baser impulses in check most of the time through self-regulation. With the right mix of social pressure and perceptions however, these self-regulating mechanisms can be suspended in a process called moral disengagement, and unethical outcomes can follow (Bandura, 1998, 2004). Moral disengagement can be initiated in four ways: First, individuals can develop a moral justification for their unethical behavior and then recast the unethical action as a moral obligation. This justification can then be reinforced by making favorable comparisons with others who have behaved worse in similar situations. Then, by employing euphemistic labels – "collateral damage" and "neutralizing the enemy" are two military examples – individuals can avoid thinking about the moral implications of their role and the original moral justification can stand. Second, they may deflect or displace responsibility for any harm they cause, or diffuse the responsibility throughout their unit. Third, they can misrepresent any destructive consequences which might result from their decisions by distorting, discrediting or ignoring any harmful results. Fourth, they may dehumanize their victims. It is easier to abdicate responsibility for harming others if we can first strip victims of their humanity or blame them for their plight. Again, the use of labels, pejorative terms and symbols are important ingredients in dehumanization because they help create distance between the perpetrator and the victim. It is noteworthy that the use of symbols and labels has been identified by Stanton (1996) as the second step in his eight-stage model of genocide. It is also common for soldiers to use labels when referring to out-group members. For example, soldiers of the above-mentioned Canadian Airborne Regiment used terms like "slomali," "smufty," and "nignog," when referring to Somali nationals "although many ... said they did not think of them as racist epithets" (Canada, DND, 1997, 293). Of course, the use of labels and derogatory terms by soldiers will not necessarily lead to unethical behavior, but it can, and therefore needs to be closely monitored by military leaders.

Let me sum up the argument made so far. Individuals are prone to errors in judgment, and judgment is one of the more important attributes associated with moral behavior. In the course of daily life, individuals are bombarded by external influences—some accompanied by ethical risks—emanating from society, the institutions they belong to, and the immediate group of people with whom they live and work. These influences are not always harmful, however. External influences can be positive and lead to personal growth. The military, in particular, has been a source of great personal development for many. That said, the aim of this chapter is to highlight those aspects of military life that can challenge the ethical competence of individuals, particularly individuals who do not possess in strong measure the moral attributes identified earlier (moral sensitivity, judgment, motivation, character, empathy, locus of control, responsibility, and agency). The remainder of this chapter examines the ethical challenges inherent in military culture in greater detail.

Ethical Risks inherent in Military Life

While there is much about military life that is positive and contributes to the personal and moral development of soldiers, there are three potential problems that military culture poses for ethical decision making. First, the underlying values that guide much of military thought and behavior may not be conducive to moral reflection and autonomy, which are essential elements for moral functioning. Second, the military's emphasis on rapid decisions and action may lead military personnel to approach moral problems with the same hasty approach instead of thinking deeply about the underlying issues. Third, it is possible that aspects of military culture which reinforce the hierarchical nature of the military may encourage individual soldiers to abdicate their moral responsibility.

Ethics and the Military Mindset

Over the past decade or more, the military forces of most western nations have established ethics training programs (Robinson, de Lee and Don Carrick, 2007; Toiskallio, 2007). Most of these programs advertise a list of military values or virtues defining what it means to be a professional military man or woman in that nation (see Robinson [2007] for a list of the values of the US Army, British Army, and Canadian Forces). Of course, these are espoused values, the values a military force would like to see in its personnel. But what values are actually practiced? Unfortunately, this is an open question as there is no empirical research on this subject beyond the occasional generalizations which emerge when transgressions of military personnel are made public. The extent to which espoused military values correspond to actual military practice is an important line of research that should be taken up by scholars because the current literature does not paint the military as an environment conducive to ethical competence. For example, Huntington (1957, 60–1) described the military mindset as conservative, anti-individualistic and pessimistic in the sense that it viewed human nature as a mix of strength and weakness, with weakness being the more dominant of the two. More recently, Wolfendale (2007, 128) has suggested that military culture, particularly culture in military training settings, promotes a totalitarian, authoritarian, anti-individualistic, hierarchical, and conservative mindset. With such qualities, the military hardly seems the sort of environment that will encourage individuals to engage in moral reflection and to consider the interests of others, concepts usually associated with ethical functioning.

Military culture will certainly vary from nation to nation, and also within the services and units of a particular nation. In addition, there are many ethical leaders and ethical soldiers in the military, but with little empirical research to support this assertion, the stereotype of military culture as incompatible with ethical growth is permitted to stand. The situation is further aggravated when we continue to see incidents of soldiers behaving unethically, and are able to trace these ethical failures to problems with military leadership and culture. Although most soldiers appear to be morally competent, further empirical research is needed to support this position.

Moral Reflection

Notwithstanding the research on intuitive decision making mentioned earlier (Haidt, 2001), the popular belief is that moral reflection is an important precursor to moral action. Opportunities for personal reflection are now an important part of the education programs in many professions (Lyons, 2010). Educators contend that giving people the analytical tools for reflecting on moral problems and creating an environment that supports reflection and moral autonomy is necessary for enhanced moral functioning. But how far can this idea go in the military?

Although we may prefer to have military personnel who engage in reflective thinking, there are aspects of military culture that encourage a more rapid style of automatic thinking that is antithetical to reflective thought. First, the need for quick and decisive action in military operations can persuade military personnel that quick decisions and rapid responses are valued over reflective thought in other areas as well. While quick decisions are often required in combat, military personnel are not in combat most of the time. The value placed on rapid decision making can be carried to other types of problems where there is more time and perhaps a greater need to reflect on decisions, as in situations where the agent has little or no experience. In the beating death of the young Somalia man by Canadian paratroopers mentioned earlier, the abuse lasted over a period of hours and a number of soldiers could have stopped it at any time before the youngster died. But they didn't, perhaps because they were unable, or unwilling, to reflect on the moral dimensions of the situation and recognize their professional obligation to intervene, or because they automatically decided that it was someone else's problem.

Second, the military's emphasis on rule-based decision making can result in people restricting their analysis to a search for the correct rule to apply rather than engaging in any deep reflection on the issues at stake (Dundon, 2008). Aids like standard operating procedures, doctrine, and checklists are quite helpful in preparing military personnel to respond appropriately when faced with the uncertainty of novel experiences or the stresses of combat, but these same conventions routinize work and can lead military personnel to focus more on following rules than reflecting on the moral dimensions of the tasks at hand.

Third, the military is a mission-oriented institution that emphasizes the importance of achieving goals, so there is a possibility that people may place greater value on achieving the outcome than on the means taken to achieve the desired ends. Once a mission has been declared and orders issued, it is relatively easy for soldiers to focus mostly on the objective and to rationalize their actions as moving the unit closer to achieving the mission. At Abu Ghraib, for example, senior leaders pressed their juniors to obtain actionable intelligence from detainees, so that soldiers' lives could be protected (Danner, 2004, 20). For some of those involved in this incident, the importance of this mission overrode any consideration of the morality of how they were treating the detainees.

Fourth, the way work is structured in the military can also lead to unethical behavior. Military work is typically compartimentalized and segmented into specialized roles, particularly at the lower ranks, so that individuals will work on only a small portion of the broader mission assigned to their unit. At the same time, information is controlled so that it is withheld from some and parcelled out only to those who "need to know," making it possible for personnel to miss the moral implications of the larger task of which they are a part and to ignore their responsibility. Managing military jobs in this way certainly saves time and training resources, but it can also narrow the soldier's perspective to the point where he or she may lose sight of the broader moral implications of the mission they are involved in. Imagine, for example, a soldier who has detained someone and then passed the detainee on to another unit, where the detainee is then tortured. In such a segmented chain of tasks and responsibilities, it is not difficult for the soldier who initially detained the suspect to overlook his or her role in the detainee's suffering at the hands of those further along the chain. In another example, Browning (1998, 162) found that breaking work down into specialized functions was one of the factors that helped World War II German police abdicate their personal responsibility as they went about the job of rounding up and killing Jews in Poland in 1942. Similarly, the segmenting of jobs and routinization of procedures were identified as contributing factors in the My Lai massacre of 1968 (Kelman and Hamilton, 1989, 18). (For a philosophical analysis of the impact that compartimentalizing work has on moral functioning see MacIntyre [1999]).

Fifth, while military doctrine and policy emphasize loyalty to professional military values and precepts, the practical reality of day-to-day activities in military units reinforces loyalty to one's peers, one's superiors, and one's subordinates. In this way, loyalty is personalized and when loyalty to one's peers or superiors comes into conflict with one's professional obligations to higher-order principles, not every solder has the time, training or ability to reflect on the moral obligations implicit in the choices at hand.

Moral Autonomy and Agency

Another important ingredient of ethical functioning is the extent to which people have the discretion to reflect on moral issues, make choices and then follow through with moral action. Military forces have many rules and regulations to guide decision making, but there are times when rule-compliance will not lead to ethically-correct choices. There are many examples in military history where commanders have felt a need to disobey orders for ethical reasons (Rescher, 1990; Mantle, 2007) and there are a number of scholars in the military community who have argued that there are conditions which justify active disobedience (Huntington, 1957, 74–8; Green, 1985, 72; Osiel, 2002, 315).

If we expect soldiers to do the right thing regardless of regulations and directions from superiors, they need to be encouraged to exercise moral agency and taught when it is correct to choose their own judgment over rules and regulations. This means a greater sense of moral autonomy for the individual. Susan Martinelli-Fernandes (2006) has recently suggested that moral autonomy should be one of the objectives of military ethics training and education. Taken from the work of German philosopher Immanuel Kant, the concept of moral autonomy refers to the ability of an individual to derive moral principles and to act in accordance with them.

This begs the question: How much moral autonomy can be tolerated in the military domain? After all, military forces are by nature hierarchical—direction comes from the top. As a result, military culture has always had a certain tolerance for moral autonomy where senior leaders are concerned, but less so in the case of junior leaders. On the other hand, a certain degree of flexibility is occasionally permitted by commanders to meet the unexpected demands of the tactical situation and to capitalize on the talents of leaders on the ground (Osiel, 2002, 316), but the military is a bureaucratic profession and junior leaders generally do not understand the big picture to the extent that senior leaders do. In the end, it may be unrealistic to expect, or undesirable to want, too much in the way of moral autonomy on the part of junior leaders.

Developing moral autonomy in the military will be difficult. Timothy Challans, a former military ethics educator, has challenged the US military's system of teaching ethics to its members as rooted in authority, dogma and indoctrination. He suggests that the US military ethos needs to change from an orientation based on moral heteronomy (relying on direction from another person, text or law) to moral autonomy (based on freely-chosen principles grounded in reason), with emphasis on "moral autonomy instead of moral authority, public reason instead of ideology, and moral justification instead of foundation (Challans, 2007, 184).

Full moral autonomy throughout the chain of command may be too much flexibility for the military to bear. However, there is potential for a certain degree of moral autonomy for those cases in which soldiers are placed in morally ambiguous circumstances, whether through the orders of morally incompetent leaders or simply because of unanticipated events. In this regard, moral autonomy is like judgment and initiative, two attributes that are valued in military operations. If good judgment and initiative are to be admired in tactical matters, these attributes should also be encouraged in moral matters. Unfortunately, several aspects of military culture seem to hinder moral autonomy and promote instead a tendency towards moral deference to others.

First, the emphasis on cohesion and loyalty in the military, as mentioned earlier, can lead to the deindividuation of soldiers to the point that they lose their moral identity in the group and conform to situational demands instead of using their own moral reflection and agency. Several aspects of military life contribute to deindividuation, which in turn hinders moral autonomy and agency. First, the social processes employed to enhance unit cohesion can overwhelm individual identity (e.g., parade square drill, unit chants, and intensive physical training), which in turn can lead to a sense of anonymity, subordination to the will of the group and unethical behavior if situational demands call for such behavior. Second, it is possible that simply wearing military uniforms and personal protection gear (helmets, flak jackets, respirators, etc.), which obscure the soldier's identity, also aid the process of deindividuation, because they make all soldiers look alike and contribute to anonymity. Because military units have powerful norms and strong cultures (i.e., where consensus is clear on what is right and wrong), it is relatively easy for soldiers – particularly soldiers with low self-esteem or lack of self-confidence – to deactivate their self-control mechanisms (i.e., to morally disengage) and follow the influence of the group. As long as the influence of the group is aimed at ethical ends this is not a problem, but it becomes a problem when the group has unethical intentions.

Second, the hierarchical nature of the military might lead soldiers' to exercise moral deference when they should be exercising moral agency. Aspects of military culture that promote authority and respect for the chain of command can also contribute to moral deference, particularly in cases where individuals do not have a strong moral identity of their own. Like deindividuation discussed above, elements like the symbols of military rank and authority, standard operating procedures, charismatic leaders, the segmentation of military jobs, and command-centric operations all encourage soldiers to defer to those at the higher levels of command. In command-centric operations, for example, where junior personnel are busily passing information to the commander for him or her to make decisions, it is natural for junior personnel to think that the commander understands the big picture, will make the big decision, and is therefore responsible for the moral implications of any consequences. Most of the time this does not pose any problem, but in those cases where orders are unethical or ethically ambiguous, soldiers can be lulled into accepting without reflection the moral vision painted by an influential leader.

Conclusion: Where Do We Go From Here?

The aim of this chapter has been to present a psychological examination of the forces that pose ethical challenges for military personnel. Some of these influences can be mitigated by senior leaders who will create ethical climates in their units and quality training that can teach personnel how to exercise moral competence in difficult situations. In today's operations, however, where the individual soldier can have strategic implications (Krulak, 1999), the moral functioning of military forces in the field will be determined in large measure by junior personnel. If we want soldiers to behave professionally in the face of these risks, we must prepare them for the difficult environments they will work in. I will close this chapter with four thoughts on how we might proceed.

First, develop the moral competence of the individual soldier. Military training should focus on inculcating in junior personnel and aspiring leaders the personal attributes needed to take proper moral action when required. This development should involve training aimed at enhancing moral awareness and judgment, such as teaching soldiers how to conduct ethical analyses without succumbing to decision-making errors, and then educating them about the social-psychological influences that might derail their moral judgment and motivation. Had the soldiers of the Canadian Airborne Regiment been prepared along these lines, they would have been aware of the environmental and group factors that could possibly influence them while on operations in Somalia, and they would have been better able to take proper moral action.

Second, empower the soldier so he or she will take moral action. Soldiers should be taught how to take moral action when necessary. This may include instruction on how to raise matters with the chain of command, or to take matters outside the chain of command when necessary, perhaps via ethics hot-lines or inspector-general channels.

Third, publicize transgressions. Military forces are very good at learning from tactical mistakes on the battlefield. This adaptive quality needs to be carried over to the moral domain. Moral failures should be analysed, publicized, discussed and used as teaching tools in professional military training and education. Unfortunately, there is a tendency in military organizations to move quickly past incidents of ethical or professional failure, with the result that individual members are denied the opportunity to digest the lessons to be learned from such events.

Fourth, measure ethical climate. If you want to change behavior, measure it. Measuring the ethical climate of units on a regular basis will convey to commanders that superiors are serious about ethics. There are instruments for measuring ethical climate (Victor and Cullen, 1988) and some have been adapted for military use. For example, the CF has conducted major surveys of ethical climate in recent years (Catano, Kelloway, and Adams-Roy, 2000; Dursun, Morrow, and Beauchamp, 2004; Fraser, 2008) and the Land Force (i.e., Canadian Army) regularly measures the moral climate of units deployed on operations.

Military operations challenge the moral functioning of everyone involved. In addition, military organizations and procedures have inherent qualities that can lead individuals to make unethical choices. Consequently, military leaders and educators alike have a responsibility to prepare junior soldiers and aspiring leaders for the ethical challenges ahead. I hope this chapter has provided some insights that can help with this important work.

Bibliography

Aronson, E. (2004), *The Social Animal* (9th Edition) (New York: Worth Publishers).

Bandura, A. (1998), "Mechanisms of Moral Disengagement," in W. Reich (ed.), *Origins of Terrorism: Psychologies, Ideologies, Theologies, States of Mind* (New York: Cambridge University Press).

Bandura, A. (2004), "The Role of Selective Moral Disengagement in Terrorism and Counterterrorism," in M. Fathali Moghaddam and A.J. Marsella (eds), *Understanding Terrorism* (Washington DC: American Psychological Association).

Bebeau, M.J. (2002), "The Defining Issues Test and the Four Component Model: Contributions to Professional Education," *Journal of Moral Education*, 31:3, 271–95.

Ben-Dor, G. and Pedahzur, A. (2004), "The Uniqueness of Islamic Fundamentalism and the Fourth Wave of International Terrorism," in L. Weinberg and A. Pedahzur (eds), *Religious Fundamentalism and Political Extremism* (London: Frank Cass Publishers).

Bercuson, D. (1996), *Significant Incident: Canada's Army, the Airborne, and the Murder in Somalia* (Toronto: McClelland and Stewart).

Bilton, M. and Sim, K. (1992), *Four Hours in My Lai* (New York: Penguin).

Browning, C. (1998), *Ordinary Men: Reserve Police Battalion 101 and the Final Solution in Poland* (Harper Perennial).

Canada, Department of National Defence. (1997), *Dishonoured Legacy: The Lessons of the Somalia Affair (Report of the Inquiry into the Deployment of Canadian Forces to Somalia)*, Vol. 1.

Canada, Department of National Defence (2001), *Debrief the Leaders Project (Officers)*.

Catano, V., Kelloway, K. and Adams-Roy, J.E. (2000), *Measuring Ethical Values in the Department of National Defence: Results of the 1999 Research* (Sponsor Research Report 2000–21).

Challans, T.L. (2007), *Awakening Warrior: Revolution in the Ethics of Warfare* (Albany, NY: State University of New York Press).

Colby, A. (2008), "Fostering the Moral and Civic Development of College Students," in L.P. Nucci and D. Narvaez (eds), *Handbook of Moral and Character Education* (New York and London: Routledge).

Damasio, A. (1994), *Descartes' Error: Emotion, Reason, and the Human Brain* (New York: Penguin).

Danner, M. (2004), *Torture and Truth: America, Abu Ghraib, and the War on Terror* (New York: The New York Review of Books).

Dovidio, J.F., Piliavin, J.A., Schroeder, D.A. and Penner, L.A. (2006), *The Social Psychology of Prosocial Behavior* (Mahwah, NJ: Lawrence Erlbaum Associates).

Dundon, R. (2008), *Manipulation of Moral Agency by Western Armed Forces,* paper presented at the War Studies Research Day, May 14, 2008 (Kingston, Ontario, Canada: The Royal Military College of Canada).

Dursun, S., Morrow, R.O., and Beauchamp, D.L.J. (2004), *2003 Defence Ethics Survey Report* (Sponsor Research Report 2004–18).

Forte, A. (2005), "Locus of Control and the Moral Reasoning of Managers," *Journal of Business Ethics*, 58:1, 65–77.

Fraser, K. (2008), *The 2008 Defence Ethics Survey Analysis: Findings for the Canadian Forces and the Department of National Defence* (Defence Research and Development Canada – Toronto, CR 2009 – 196).

Gladwell, M. (2005), *Blink: The Power of Thinking Without Thinking* (New York: Little, Brown and Company).

Green, L.C. (1985), *Chapters on the Modern Law of War* (Dobbs Ferry, NY: Transnational Publishers).

Haidt, J. (2001), "The Emotional Dog and Its Rational Tail: A Social Intuitionist Approach to Moral Judgment," *Psychological Review,* 108:4, 814–34.

Hauser, M.D. (2006), *Moral Minds: The Nature of Right and Wrong* (New York: Harper Perennial).

Hintjens, H. (June 1999), "Explaining the 1994 Genocide in Rwanda," *The Journal of Modern African Studies*, 37:2, 241–86.

Horn, B. (2001), *Bastard Sons* (St. Catherines, Ontario: Vanwell Publishing).

Horn, B. (2007), "What did you expect? An Examination of Disobedience in the Former Canadian Airborne Regiment, 1968–1995," in H.G. Coombs (ed.), *The Insubordinate and the Noncompliant* (Kingston, Ontario, Canada: Canadian Defence Academy Press, and Toronto: The Dundurn Group).

Huntington, S.P. (1957), *The Soldier and The State: The Theory and Politics of Civil-Military Relations* (Cambridge, MA: Belknap Press).

Janis, I.L. (1972), *Victims of Groupthink* (Boston: Houghton Mifflin).

Kamenetsky, C. (April–June, 1997), "Folktale and Ideology in the Third Reich," *The Journal of American Folklore*, 90:356, 168–8.

Kelman, H.C. and Hamilton, V.L. (1989), *Crimes of Obedience: Towards a Social Psychology of Authority and Responsibility* (New Haven, CT: Yale University Press).

Kohlberg, L. (1976), "Moral Stages and Moralization: The Cognitive-Developmental Approach," in T. Lickona (ed.), *Moral Development and Behavior: Theory, Research and Social Issues* (New York: Holt, Rinehart, and Winston).

Krulak, C.C. (1999), "The Strategic Corporal: Leadership in the Three Block War," *Marines Magazine*, available online at: http://www.au.af.mil/au/awc/awcgate/usmc/strategic_corporal.htm.

Lyons, N. (2010), *Handbook of Reflection and Reflective Inquiry* (Springer).

MacIntyre, A. (1999), "Social Structures and their Threats to Moral Agency," *Philosophy*, 74:289, 311–29.

Mantle, C.L. (2007), *The Apathetic and the Defiant: Case Studies of Canadian Mutiny and Disobedience*, 1812–1919 (Kingston, Ontario, Canada: Canadian Defence Academy Press).

Manning, F.J. (1991), "Morale, Cohesion, and Esprit de Corps," in D.A. Mangelsdorff and R. Gal (eds), *Handbook of Military Psychology* (London: John Wiley).

Marcus, G. (2008), *Kluge: The Haphazard Construction of the Human Mind* (New York: Houghton Mifflin Company).

Martinelli-Fernandez, S. (2006), "Educating Honorable Warriors," *Journal of Military Ethics*, 5:1, 55–66.

Merkhoffer, L. (no date), available online at: http://www.prioritysystem.com. reasons1.html.

Milgram, S. (2004), *Obedience to Authority: An Experimental View* (New York: Perennial Classics).

Narvaez, D. and Rest, J.R. (1995), "The Four Components of Acting Morally," in W.M. Kurtines and J.L. Gewertz (eds), *Moral Development: An Introduction* (Boston, MA: Allyn and Bacon).

Osiel, M.J. (2002), *Obeying Orders: Atrocity, Military Discipline and the Law of War* (New Brunswick, NJ: Transaction Publishers).

Pascarella, E.P. and Terenzini, P.T. (1991), *How College Affects Students: Findings and Insights from Twenty Years of Research* (San Francisco: Jossey-Bass).

Pascarella, E.P. and Terenzini, P.T. (2005), *How College Affects Students: A Third Decade of Research Vol. 2* (San Francisco: Jossey-Bass).

Rescher, N. (1990), *In the Line of Duty: The Complexity of Military Obligation*, The Joseph A. Reich, Sr., Distinguished Lecture on War, Morality, and the Military Profession, presented at the United States Air Force Academy, Colorado, 15 November 1990.

Rest, J. (1986), *Moral Development* (New York: Praeger Publishers).

Rest, J.R. (1994), "Background: Theory and Research," in J.R. Rest and D. Narvaez (eds), *Moral Development in the Professions: Psychology and Applied Ethics* (Hillsdale, NJ: Lawrence Erlbaum Associates).

Robinson, P. (Spring, 2007), "Ethics Training and Development in the Military," *Parameters*, 37:1, 22–36.

Robinson, P., de Lee, N. and Carrick, D. (2008), *Ethics Education in the Military* (Aldershot: Ashgate).

Shaw, M. (2005), *The Western Way of War: Risk-Transfer War and its Crisis in Iraq* (Cambridge: Polity Press).

Shorey, G. (Winter 2000–2001), "Bystander Non-Intervention and the Somalia Incident," *Canadian Military Journal*, 1:4, 19–28.

Shorey, G. (2006), "Disobedience of Professional Norms: Ethos, Responsibility Orientation and Somalia," in C.L. Mantle (ed.), *The Unwilling and The Reluctant: Theoretical Perspectives on Disobedience in the Military* (Kingston, Ontario, Canada: Canadian Defence Academy Press).

Smith, T.W. (2008), "Protecting Civilians…or Soldiers? Humanitarian Law and the Economy of Risk in Iraq," *International Studies Perspectives*, 9, 144–64.

Stanford Encyclopedia of Philosophy, available online at http://plato.stanford.edu/enries/moral-dilemmas/.

Stanton, G.H. (1996), *The 8 Stages of Genocide*, originally presented as a briefing paper at the U.S. State Department, available online at: http://www.unitar.org/ny/sites/default/files/Eight%20Stages%20of%20Genocide.pdf.

Streatfield, D. (2007), *Brainwash: The Secret History of Mind Control* (New York: St. Martin's Press).

Tavris, C. and Aronson E. (2007), *Mistakes Were Made (but Not By Me): Why We Justify Foolish Beliefs, Bad Decisions, and Hurtful Acts* (Orlando, FL: Harcourt Books).

Thoma, S.J. (2002), "An Overview of the Minnesota Approach to Research in Moral Development," *Journal of Moral Education*, 31:3, 225–45.

Toiskallio, J. (2007), *Higher Education in the Military: What, How and Why in the 21st Century* (Helsinki: Finnish National Defence University).

Trevino, L.K., Weaver, G.R. and Reynolds, S.J. (2006), "Behavioral Ethics in Organizations: A Review," *Journal of Management*, 32:6, 951–90.

Vaughan, D. (1999), "The Dark Side of Organizations: Mistake, Misconduct, and Disaster," *Annual Review of Sociology*, 25, 271–305.

Victor, B. and Cullen, J.B. (1988), "The Organizational Bases of Ethical Work Climates," *Administrative Science Quarterly*, 33, 101–25.

Walker, L.J. (2004), "What Does Moral Functioning Entail?" in T.A. Thorkildsen and H.J. Walberg (eds), *Nurturing Morality* (New York: Kluwer Academic/Plenum Publishers).

Winslow, D. (2004), "Misplaced Loyalties: Military Culture in the Breakdown of Discipline in Two Peace Operations," *Journal of Military and Strategic Studies* 6:3, 1–18.

Wolfendale, J. (2007), *Torture and the Military Profession* (Basingstoke: Palgrave Macmillan).

Zimbardo, P. (2007), *The Lucifer Effect: Understanding How Good People Turn Evil* (New York: Random House).

Chapter 13

Moral Formation of the Strategic Corporal

Rebecca J. Johnson

A platoon leader interviewed in Iraq in 2004 explained the challenge he faced leading his soldiers in stability operations:

> It is very difficult to keep 18-year-old guys, to take them and one second we are dodging bullets and trying to hide on the street corner and react because somebody got somebody in a window or a roof, and the next second you are knocking on the door, asking to search the house and you have to be polite. I think that is a very large leadership challenge here—keeping guys focused on that; making sure that they can calm back down after brief periods of excitement. (quoted in Wong 2004, 5)

This lieutenant identifies the difficulty at the tactical level of irregular warfare noted by former Marine Commandant Charles Krulak in "The Three Block War." In contemporary wars, junior ground forces find themselves "fighting like the dickens on one block ... handing out humanitarian supplies on the second block, and the next one over ... trying to keep warring factions apart" (Mattis and Hoffman 2005, 19).

Success at this type of warfare demands the development of what General Krulak has named the "strategic corporal"—a junior Marine (or soldier) who possesses the capacity, flexibility, and resilience to demonstrate accurate situational awareness; calculate the tactical, operational, and strategic first, second, and third order effects of potential courses of action; interpret commander's intent and any operational orders (OPORDS) or fragment orders (FRAGOs) in light of these assessments; and choose and execute an appropriate course of action. The strategic corporal will range in age from roughly 22–24 and will have been a member of the Armed Services for a little less than four years. A little over five percent will be on their first combat deployments.

The officer leading the strategic corporal's platoon will likely be under 25 years old and on his first combat tour as well. The infantry squad leader (an E5 sergeant in the Marines) will be between 24 and 27 and will have been in the Corps about six years. The infantry platoon sergeant/company gunnery sergeant is a 29–34-year-old with roughly 10 to 14 years of service.

Given these realities, what enables junior soldiers and Marines to make morally sound decisions in irregular war? This chapter argues that moral action in irregular warfare requires a level of moral sensitivity and intuitive judgment that is not supported by current military training and educational initiatives. What is more, external incentives motivate junior soldiers and Marines to behave in a way that is more aggressive than is appropriate for irregular warfare. The erosion of small unit cohesion exacerbates these problems, leaving junior troops vulnerable to committing abuse and suffering significant psychological trauma.

This chapter will explore these issues in four sections. The next section will identify the unique moral challenges faced by junior soldiers and Marines in irregular warfare. Small units operate in a battlefield that includes an ever-growing number of relevant actors and missions and arrive at their deployment often lacking the cohesion needed to operate as an effective fighting team; yet such groups are responsible for acting with minimal guidance and maximum discretion. The second section will outline the significance of these changes, which can undermine the potential for overall mission success and strategic victory, increase the potential for abuse, and threaten the psychological and social wellbeing of individual soldiers and Marines. As the Army and Marine Corps embrace the strategy of sending small units into dangerous areas with minimal support, this significance strengthens. Section 3 draws from James Rest's model for moral behavior to explain how its four components—moral sensitivity, judgment, motivation, and character—relate to irregular warfare and how the model can be applied to the moral development of junior soldiers and Marines. The conclusion offers specific recommendations for how to utilize Rest's model to prepare junior ground forces to excel morally in irregular warfare.

Moral Decisionmaking in Irregular Warfare: A Whole New World?

The conventional idea of war as defined by combatants relying on overwhelming force to erode the enemy's will and/or capacity to fight is challenged by irregular war. The use of force is certainly one essential component, but irregular warfare is characterized by an increased number of actors in the "battlespace" who use a range of lethal and non-lethal means to complete military, political, economic, and social missions. The increased complexity of the unconventional environment makes it harder to discriminate between combatants and noncombatants and anticipate the effects of different courses of action. Likewise, the complexity of irregular warfare requires those "on the ground" to decide whether and when to escalate force. Both of these complexities imply that decision making authority devolves to junior leaders out conducting missions. While these individuals have the best vantage point on a given situation, that position does not always provide clarity. Some 31 percent of Marines and 28 percent of soldiers in Iraq in 2005–2006 reported encountering ethical situations to which they did not know how to respond (MHAT IV, 2006, 37). This places extraordinary responsibility on relatively young and inexperienced soldiers and Marines at the same time that a growing lack of unit cohesion strains the junior leader's ability to guide and mentor his or her subordinates. This section will examine these dynamics in an effort to clarify how today's "strategic corporal" faces moral challenges of a special kind.

Proliferation of Actors and Missions Creates a Complex Moral Landscape

In conventional war, the decision of when and how to use lethal force is guided by two principles—the principles of discrimination and proportionality. Stated simply, combatants must discriminate between combatants and noncombatants, may target non-prohibited weapons against other combatants only, and may use those weapons only to the extent that the harm caused does not exceed the good achieved.

As anyone with a passing familiarity with counterinsurgency knows, the discrimination necessary to preserve noncombatant immunity is hard to achieve because insurgents routinely hide among the population as means of cover and support. They do not wear uniforms or a "fixed distinctive sign recognizable at a distance." They do not carry their weapons openly. They do not abide by the laws or customs of war and thus often launch attacks out of civilian centers to draw return fire on noncombatants. As a result, when moving in convoys or conducting street operations, ground forces can understandably have a hard time identifying combatants. Troops have devised ingenious ways to identify potential combatants; for example, low-riding cars or cars with new tires are immediately identified as potentially holding explosives and their drivers are targeted. In contrast, drivers who ignores Marines' call to stop in deference to a passing convoy may not be targeted because Iraqi drivers have a reputation for low situational awareness behind the wheel and Iraqi cars have a reputation for bad brakes. These "hunches", however, do not constitute actual discrimination.

Forces face a similar difficulty with proportionality (Wong 2004, 14). The calculation of proportionality (that the military benefit of any action outweighs the harm caused by the action) is challenging enough given the fog of war. In irregular war troops must calculate the direct effects of achieving a military objective; for instance, can they conduct a mission in pursuit of an insurgent who has just targeted a coalition convoy given the risk the mission poses to civilians in the village? They must also calculate the second and third-order effects of their mission. These indirect effects include such issues as whether the mission would breed ill will among the villagers, perhaps pushing some civilians to join the insurgency. Would it undermine ongoing efforts of other actors to cultivate restraint within the existing law enforcement community? Does it hand a public relations victory to the enemy by providing images of soldiers and Marines standing over villagers cowering in the road—a significant offense in Iraqi culture? Calculating these direct and indirect effects to determine proportionality becomes unbelievably challenging (Richardson et al., 2004, 104).

In such a complex environment, decisions about proportionality necessarily devolve to the service member in charge of the mission to clear the village. The commanders back on the base lack the situational knowledge to make the call.

Only those in the position to pursue the insurgents into the village, ask the village elder for permission to search the village, ask the elder to apprehend the insurgents himself, or retreat entirely can decide whether the actions would be proportionate (Richardson et al., 2004, 104).

In military doctrine, this concept is identified as "mission command:" commanders identify their intent for specific missions, and subordinates use their discretion and judgment in accomplishing mission objectives. Tactically this makes good sense—they are the soldiers and Marines closest to the ground and in the best position to assess an area's needs and appropriate response. At the same time, while versed in the law of war and rules of engagement, most junior soldiers and Marines lack the experience and perspective of their more senior counterparts that helps ground sound judgment and guard against abuse. It is a result of the devolution of operational authority that soldiers and Marines at lower levels are increasingly responsible for making sophisticated tactical judgments that carry complex moral ramifications (Wong 2004, 14).

Lack of Unit Cohesion Inhibits Moral Action in Irregular Warfare

As General Krulak notes, "The clear lesson of our past is that success in combat, and in the barracks for that matter, rests with our most junior leaders" (1999). Small group leadership is key. Successful small unit leadership rests on eleven principles: gaining technical and tactical proficiency; demonstrating self-awareness and a commitment to self-improvement; possessing knowledge of one's soldiers or Marines along with a commitment to their welfare; maintaining clear communication with one's unit; leading by example; ensuring task are understood, supervised, and accomplished; training their soldiers and Marines as a team; making sound and timely decisions; developing a sense of responsibility among subordinates; employing one's unit in accordance with its capabilities; and pursuing responsibility, while accepting responsibility for one's actions (FMFM 1–0, 1995, 105; Malone 1983, 32–4).

While intuitive, these principles take time to form a cohesive small unit, even with the most talented leader. Unfortunately, small unit leadership has been eroded by increased operational tempo and a reduced culture of mentorship. To the extent that the relationship between the squad leader and his troops is weakened, even if he possesses the sophistication needed to make morally complex decisions quickly in situations of significant stress, it cannot be assumed that his troops will possess the same ability. For example, his troops may lag behind him developmentally and/or they may possess personality types that struggle with complexity and uncertainty (Kohlberg 1981; Zuckerman 1979; Petrie 1967; Breivik 1999). According to developmental psychologist Robert Kegan, most young adults in their late teens and early twenties demonstrate a propensity to define their actions by their peer group rather than by internal constraints or directed rules (Kegan 1998). Individuals tend to evaluate behavior self-referentially by asking what the group values as good for the individual. This means that individuals within a tight-knit peer group that models discipline will tend to demonstrate discipline, while individuals within a loose-knit peer group that models a lack of discipline will demonstrate the same.

The implications for moral behavior are significant. Strong group cohesion does not necessarily compel moral behavior; clear evidence exists of group conformity motivating immoral behavior (Milgram 1974; Muñoz-Rojas and Frésard 2004, 6). Strong group cohesion partnered with a belief in the legitimacy of organizational authority is necessary to constrain group behavior ethically (Tyler et al., 2007).

This works well in military training when unit cohesion can reinforce specific patterns of behavior that would otherwise seem absurd, such as putting oneself in harm's way to complete a mission or demonstrate restraint when under threat (Forsythe et al., 2005). When unit cohesion is low, this value on self-sacrifice and discipline grows weaker as the value placed on the unit as the members' central defining group weakens. Other social groups may remain primary in the minds of the squad or platoon members and it is their values, not the small unit values, which will shape the individuals' ideas of right and wrong. The 2006 Army Mental Health Assessment Team survey dealing with battlefield ethics (MHAT IV) found this to be true as well, noting that " ... Soldiers and Marines who reported better officer leadership were more likely to follow the ROE than those Soldiers and Marines who reported poorer officer leadership" (MHAT IV, 2006, 35). Strong leadership correlates with ethical behavior; poor leadership correlates with ethical lapses.

Finally, fractured unit cohesion is especially dangerous when those units lose members. If the squad and platoon leaders are unable to manage the very real rage that results when a unit member is killed or injured, they are setting their units up to express that rage in destructive, counterproductive, and likely immoral ways. Unit cohesion cannot lessen the pain or rage felt when a friend is hurt, but it can help strengthen the individual members' resolve to exert the self-control necessary to prevent abuse (see MHAT IV for the effect of unit casualties on noncombatant abuse. MHAT IV, 2006, 41; Osiel, 1999, 188; Rosen 2005, 121). The Marine tenet of "leadership by example" is critical here. Obviously the unit leader must model appropriate restraint for his subordinates, but unless those subordinates already view him as morally authoritative, his action and guidance will have little direct effect in moments of crisis.

While it may be challenging for junior soldiers and Marines to fulfill their moral responsibilities in irregular warfare, it is critical that they do so. The ability to operate with little guidance matters for achieving strategic goals, avoiding abuse, and providing for the troops' wellbeing. It is to these topics that the chapter now turns.

The Moral Relevance of the Strategic Corporal

Aside from the expectation that soldiers and Marines behave honorably at all times, regardless of circumstance, why must the "Strategic Corporal" possess the moral sophistication required to fight honorably in a nation's irregular wars? Enemy use of nondiscriminatory tactics like IEDs and suicide bombers, and tactics that clearly target noncombatants such as "night letters" and hiding among the population, indicate that the enemy has no compunction about violating internationally accepted moral and legal standards of warfare. Still, there exists a strong conviction among uniformed personnel and the political community they serve that American troops must abide by certain moral standards in combat, regardless of their enemy's tactics. There are three primary reasons that support this belief: (1) it is a necessary condition for victory; (2) it is a necessary defense of core American and military values; and (3) it is a necessary component of the social and psychological welfare of the individual soldier and Marine. Failing to develop the moral sophistication needed to operate responsibly in irregular warfare threatens defeat, the commission of grave abuses, and individual trauma.

Morality: Key to Victory

In irregular warfare, all sides are competing for the same thing—to win the support of the population. This means that the enemy seeks to divide the population from the government while the counterinsurgents seek to ally the population with a legitimate government. Former commander of NATO's International Security Assistance Force (ISAF) and US forces in Afghanistan, General Stanley McChrystal, has recognized this fact in the guidance he provided his troops in summer 2009. According to the Tactical Directive released July 6th, "Like any insurgency, there is a struggle for the support and will of the population. Gaining and maintaining that support must be our overriding operational imperative—and the ultimate objective of every action we take" (ISAF, 2009).

Since junior soldiers and Marines are the ones patrolling villages and interacting with the population, their behavior is critical. Because of the relationship between the population and the troops, the fact that a full third of soldiers in Iraq insulted or cursed non-combatants in their presence is disturbing, even if no actual violation of non-combatant immunity occurred in those instances (MHAT V, 2008, 32).

Moral Behavior Guards against Abuse

The moral strength of the strategic corporal is important not only for winning the overall war; it is important for maintaining the moral standards of the Armed Forces by guarding against abuse. Ambiguous situations provide fertile ground for the relaxation of social standards, especially by young adults and when under threat. If Robert Kegan is correct and young adults take their social cues from their peer group, the devolution of operational authority provides a significant lack of behavioral direction to precisely the population that needs it most. Paul Bartone has identified this facet of modern warfare as at least partially responsible for the abuses at Abu Ghraib (2004, 14), as has Major General Antonio Taguba, the US Army officer responsible for investigating the abuses at the prison. His final report chronicles that "numerous incidents of sadistic, blatant, and wanton criminal abuses were inflicted on several detainees. This systematic and illegal abuse of detainees was intentionally perpetuated by several members of the military police guard force ..." (Taguba 2004, 16). The sworn statement of one sergeant indicates he witnessed behavior he found morally questionable. When asked why he did not report the behavior up his change of command, the sergeant replied, "Because I assumed that if they were doing things out of the ordinary or outside the guidelines, someone would have said something. Also the wing belongs to MI [Military Intelligence] and it appears MI personnel approved of the abuse" (Taguba 2004, 19).

One can see easily how well intentioned individuals can commit grave abuses. Research has documented multiple ways for individuals to create the type of moral distance necessary to dehumanize others (Bandura 1999; Osie, 1999). In his book, *Obeying Orders: Atrocity, Military Discipline and the Law of War*, Mark Osiel lays out the types of abuse junior soldiers and Marines are most prone to conduct—atrocity by connivance and atrocity by passion. Atrocity by connivance is the type of abuse that occurs when "troops are simply given to understand, through winks and nods of acquiescence, that spontaneously-initiated atrocities will not be penalized" (1999, 188). Osiel identified My Lai with this type of atrocity. Atrocity by passion occurs when "individual soldiers, as creatures of desire, are able to indulge their passions: for women, alcohol, food, revenge of lost comrades, or simple blood lust" (Osiel 1999, 176). Without the guidance needed to prevent both types of abuses, junior soldiers and Marines are placed in a very precarious situation.

Colonel John Mark Mattox identifies morality as a key defining feature of the military, arguing that those who defend the state from harm must be of a type different from those who threaten to do it harm—it is impossible for evil to cast out evil. The military "must themselves be of a different—and one might argue better—nature; one that not only recognizes the demands of morality but also one that consistently makes choices that reflect that recognition" (2005, 395–96). When soldiers and Marines fail to police that line between right and wrong, even when knowing the precise location of the line is difficult to impossible, they fail their overall mission, their profession and their country. Ultimately, they also fail themselves.

Morality Buffers the Trauma of War

The final reason it is important to ensure troops conducting irregular war possess the moral fortitude to make sound judgments in a dangerous, confusing environment stems from the effect that failing to do so has on the troops themselves. As described by Shannon French:

> Veterans who believe they were directly or indirectly party to immoral or dishonorable behavior (perpetrated by themselves, their comrades, or their

commanders) have the hardest time reclaiming their lives after the war is over. Such men may be tortured by persistent nightmares; may have trouble discerning a safe environment from a threatening one; may not be able to trust their friends, neighbors, family members, or government; and may have problems with alcohol, drugs, child or spousal abuse, depression, or suicidal tendencies (French 2003, 4).

The U.S. Army Mental Health Assessment Team has documented the relationship between mental health problems and unethical behavior. Their surveys indicate that soldiers and Marines who screen positive for depression, anxiety, or acute stress are twice as likely to report engaging in unethical behaviors as their counterparts who screen negative (MHAT V, 2008, 32; MHAT IV, 2006, 39). The survey does not indicate whether the unethical behavior prompted the mental health concerns or whether those individuals with existing mental health issues are more likely to behave unethically (or whether the two feed off each other in a dangerous spiral). Regardless, it is understandable why a relationship would exist between mental health concerns and unethical behavior. A class of individuals who pride themselves on their honor will suffer when forced into situations where they receive conflicting or ambiguous information concerning how to uphold it. Karl Aquino and Americus Reed identify how one's moral identity reinforces moral behavior; " ... people with a strong moral identity should strive to maintain consistency between conceptions of their moral self and their actions in the world" (2002, 1425). Built into one's moral character is an inner motivation to uphold one's self-image in action. The more central a particular moral trait is to one's self-identification, the greater lengths the person will go to protect it. The anxiety that is caused when individuals realize or come to feel they have failed themselves or their comrades in its defense is profound.

The internalization of a military ethic might also provide protection against the psychological trauma of self-betrayal by clarifying an individual's understanding of where the line between Right and Wrong lay and developing the internal fortitude required to stay of the right side of that line. According to Shannon French,

> Accepting certain constraints as a moral duty, even when it is inconvenient or inefficient to do so, allows warriors to hold on to their humanity while experiencing the horror of war—and, when the war is over, to return home and reintegrate into the society they so ably defended. Fighters who cannot say, 'This far but no farther,' who have no lines they will not cross and no atrocities from which they will shrink, may be effective. They may complete their missions, but they will do so at the loss of their humanity (French 2003, 10).

If we place young soldiers and Marines in a situation where they bear significant moral responsibility for deciding when and how to use lethal force in circumstances where it may be impossible to anticipate the effects of their actions, ensure civilians will not be killed, and meet multiple—at times competing—moral responsibilities, senior military and political leaders have an obligation to ensure they are prepared to fulfill that responsibility. If we do not prepare them adequately, their moral failing is ours, though they are the ones left to suffer the consequences.

The Elements of Moral Behavior in Irregular Warfare

If we believe that an obligation exists to help junior soldiers and Marines meet their moral obligations in irregular war, then we need to understand how. While it is essential to be able to figure out an appropriate course of action in any given situation, this ability is insufficient to guarantee moral behavior. James Rest has identified four elements that are necessary for an individual to behave morally. An individual must possess *moral sensitivity*, defined by Rest as "the awareness of how our actions affect other people" (Rest 1994, 23). Likewise, an individual must possess *moral judgment*, or the capacity to determine which course of action is (more) morally justifiable relative to the alternatives. Once individuals have determined the "right" action, they must then demonstrate *moral motivation* by prioritizing moral values over other, potentially self-serving values. Even sound prioritization is not enough, however; individuals must finally possess the *moral character* or "psychological toughness and strong character" needed to actually do the right thing (Rest 1994, 24).

Moral behavior requires all four components to function. It is simply not enough to possess highly developed powers of moral *judgment*—in irregular warfare, the ability to discriminate and calculate proportionality—if a soldier lacks the *motivation* to act morally. Likewise, the moral strength and courage required to act morally founders if it is disconnected from the moral judgment needed to guide this *character*. All of this fails if individual soldiers lacks the moral *sensitivity* to identify their actions as morally significant, for instance, by not seeing how cursing at civilians or destroying property are morally significant acts.

Rest's model identifies how the character development programs that serve as the bedrock of the Army and Marine Corps ethics training focus on moral character and judgment more than moral sensitivity and motivation. As will be demonstrated in the following pages, weakness in moral sensitivity and motivation has a profound effect on small unit action. This section will articulate each of the four components of moral behavior and their expression in irregular warfare. The following section will present suggestions for how these components may be developed in junior members of the ground forces to better prepare them for the moral challenges they face in irregular warfare.

Moral Sensitivity

The first step in moral behavior is the presence of moral sensitivity, or the ability for an individual to interpret the moral significance of a given situation and the different courses of action available in response. Moral sensitivity implies both a cognitive and an affective process; the person identifies the relevant actors and then identifies *with* them in order to evaluate how different reactions would affect them (Rest, 1994).

While people tend to focus on moral judgment as the key factor in determining moral action, judgment is impossible without moral sensitivity. In fact, Richard Hall has identified the lack of moral sensitivity as more problematic than the existence of individuals who choose consciously to behave immorally (Hall, 1975). The military often relies on the articulation of core values to serve as a method by which to identify the moral significance of a situation and interpret the benefits of different responses to it. Though each service has its own values, they focus on doing what is right, even when it is hard; placing others before oneself; and refusing to give up. These values orient the soldier or Marine to view situations in certain ways; this may help clarify ambiguous situations, but there are some limitations to relying on codes to activate moral sensitivity. First, not all moral issues are captured by these core values (Jordan 2007, 349). They provide a useful starting point, but cannot claim to be sufficiently comprehensive to direct all moral judgment. Second, while these values assist in developing moral sensitivity, they are not sufficient for developing the empathy it requires. Since these values assume the existence of a "right" that may not exist in all cultures, they stifle the ability really indentify *with* all the relevant actors. This empathy may develop as troops operate within their area and interact with the local inhabitants, but this relationship must be cultivated; it cannot be assumed.

As moral sensitivity is cultivated, individuals become increasingly proficient at making effective calculations of second and third-order effects of potential actions. This allows for a more accurate calculation of proportionality. Likewise, since soldiers and Marines are looking at a situation from the perspective of the multiple actors involved, they are better able to calibrate their actions to minimize harm and make amends when harm is done. This restraint in action and sincerity in regret can be critical to developing and maintaining popular support. Moral sensitivity is critical to moral action and it is critical to success in irregular warfare.

Moral Judgment

The second step in moral behavior is the capacity for moral judgment, or the ability to discern which actions are morally right and which are morally wrong. Moral judgment is critical, yet, it must be developed in a way that allows for rapid, intuitive decisionmaking—precisely the type of decision making that is required in irregular warfare. According to research by Jonathan Haidt and Craig Joseph, " ... human beings come equipped with an *intuitive ethics*, an innate preparedness to feel flashes of approval or disapproval toward certain patterns of events involving other human beings" (2004, 56). As moral sensitivity develops, these flashes translate themselves into action. The authors argue that over time, " ... we acquire a second nature, a refinement of our basic nature, an alteration of our automatic response" (Haidt and Joseph, 2004, 61; see also Audi, 2008). Just as soldiers and Marines are conditioned to "train like they fight" tactically (because once the reflexive reaction is built, it sticks), so are they conditioned morally.

Those, like Haidt and Joseph, who adopt an "intuitionist" approach to ethical judgment receive much opposition from those who argue that "genuine moral judgments are those that are regulated or endorsed by reflection" (Kennett and Fine 2009, 78; see also Fine 2006). While reflection certainly contributes to sound moral judgment, not every morally charged situation includes the time necessary for reflection in the moment of decision. This is where intuition, training, and mental patterns bridge the gap between prior moral reflection and present moral action.

Leonard Wong has identified the development of this moral intuition among the soldiers he interviewed in Iraq, claiming that "[b]y being confronted with complexity, unpredictability, and ambiguity, junior officers are learning to adapt, to innovate, and to operate with minimal guidance" (Wong 2004, 3). MHAT data and PTSD and suicide rates among young soldiers and Marines may challenge Wong's conclusions (MHAT V, 2008, 84–90), but the idea that soldiers and Marines develop an intuitive approach to moral judgment has much research to commend it. Burgeoning research on the effects of neuroplasticity on combat behavior suggests the brain responds to experience, creating pathways to facilitate the execution of often-repeated actions or thoughts. The brain possesses the same "muscle memory" as a basketball player's arm in conducting a jump shot or a rifleman exhaling while pulling the trigger. This is because " ... the mind, like the brain itself, is a network that gets tuned up gradually by experience. With training, the mind does a progressively better job of recognizing important patterns of input and of responding with the appropriate patterns of output" (Haidt and Joseph 2004, 62). According to Stanley et al.,

> ... with the engagement and repetition of certain mental processes, the brain becomes more efficient at those processes. Over time, the repeated engagement of the brain regions supporting a certain mental skill creates a more efficient pattern of neural activity by rearranging structural connections between brain cells involved in that skill. Thus, experience and training can lead to functional and structural reorganization of the brain. (Stanley et al., 2010)

This reorganization patterns certain responses as the brain's "default." The discrepancy between Wong's conclusion and the undeniable reality that US troops suffer the consequences of irregular warfare's moral complexity stems from the fact that adaptability is no indication that the response an individual conditions in himself is one that reinforces moral behavior. If one accepts the premise that the brain is a muscle and it is possible to develop mental muscle memory (the foundation for intuitive reactions), one must accept that muscle memory can be formed that supports immoral as well as moral reactions. Once the pattern is set, it naturally reinforces itself, becoming increasingly difficult to change.

It is also possible that muscle memory suppresses moral sensitivity, as individual soldiers and Marines disconnect themselves from the moral weight of their actions. The distancing practices that allow this suppression are well known (Muñoz-Rojas and Frésard 2004, 8–10). For example, soldiers and Marines suppress their moral sensitivity through euphemistic or dehumanizing labeling ("collateral damage", "Hajis", "bad guys", etc.), displacement of responsibility ("the insurgents hide in the villages. They're responsible for the civilians we kill when we're chasing them"), and distortion of consequences ("we treat them better than the insurgents"; "they're better off now than they were under that dictator").

The point is this: intuition is a key part of moral judgment and moral judgment relies on moral sensitivity. To the extent that an individual develops mental patterns that suppress moral sensitivity as a psychological mechanism to cope with the complexity and trauma of war, his moral judgment will suffer. He will reinforce a pattern that makes it increasingly difficult to choose moral behavior. The pre-cognitive, intuitive judgment is a given; the type of judgment is critical.

Moral Motivation

The third step in moral behavior is the existence of sufficient moral motivation, or the ability to prioritize moral values ahead of other, often-competing values. Motivation may be both internal (originating in one's own sense of Right and Wrong) and external (originating from other's incentives or penalties for behavior). Steven Silver illustrates effective external motivation in an article on Marine Corps ethics. Silver recounts the story of a Gunnery Sergeant who prevented one of his Marines from shooting an Iraqi who approached the patrol to surrender. When the Marine raised his rifle to shoot the Iraqi for his failure to stop and lay down when ordered,

> Gunny pushed it down and then dragged the Iraqi over to the line. Then he came back, got in the Marine's face, and roared, 'You don't do that s**t in my Marine Corps!" It became a running punch line. Someone was always saying, "You don't do that s**t in my Marine Corps!' and we'd all laugh. But we remembered what he said and who said it. I think it stuck. (Silver 2006, 77).

The Gunny provided his junior Marines clear behavioral guidance (they were to show restraint in the use of lethal force), backed by the highly motivating external shaming of the errant Marine in front of his platoon. The Marines took the Gunny's directive to heart and shaped their actions going forward.

One must be careful, though, to avoid confusing or competing motivations. As Paul Robinson notes in his evaluation of ethics training, "Ethics training by itself will not produce morally perfect soldiers … leaders must continually examine the health of their institutions. There is little point in teaching individuals a particular form of behavior, if they can see that the institution to which they belong in practice rewards and values other behavior" (Robinson 2007, 34). So long as soldiers and

Marines see value in receiving the Combat Action Badge or Ribbon, and they do, they will possess a strong external motivation to decide in favor of engagement over restraint. (A change in the criteria for a Combat Action Ribbon changed this somewhat; a Marine may now be eligible for the Ribbon without having returned fire, the standard before 2006.) To the extent moral development focuses on making good apples without examining the strength of the barrel, it should not be surprising when rotten apples continue to appear (Zimbardo, 2008, Chapter 15).

Moral Character

The final step in moral behavior is the possession of moral character, or the courage, persistence, focus, and agency needed to act. According to Rest, "Psychological toughness and strong character do not guarantee adequacy in any of the other components, but a certain amount of each is necessary to carry out a line of action" (1994, 24). Strength of moral character provides an individual with the flexibility needed to demonstrate the other four elements of moral behavior.

Overall, the military places special significance on individual character and works through its various training and educational opportunities to develop specific values in its members. Character is important precisely because the "fog of war" leaves individuals without clear indications of how to act. They must rely on their character for the fortitude and will to act. It is the possession of character that gives the individual soldier and Marine confidence that he will do the right thing in any given situation, regardless of whether anyone tells him what that is, and regardless of whether anyone is looking. While necessary, moral character is insufficient on its own to produce moral action. Without moral character, the other elements will likely founder. Without the other elements, moral character will be rudderless.

When all four elements come together, the result is moral action. Moral behavior takes two forms: inhibitive and proactive. The inhibitive form of moral behavior is the restraint shown when a person refrains from doing what she ought not; the proactive form is the action of a person doing what she ought. Acting proactively, people "act against what they regard as unjust or immoral even though their actions may incur heavy personal costs" (Bandura 1999, 194). The Gunny's proactive response to what he saw as a violation of the Corps' moral standards had a strong inhibitive affect on his subordinates.

How does developing junior members of the ground forces' proficiency in the four elements of moral behavior improve their ability to meet the moral challenges of irregular warfare identified in the first section? By developing a deeper appreciation for the moral contours of decisions such as where to position an outpost, whether to seek permission of village elders before searching houses, and when lethal force is justified in the pursuit of a given—known—insurgent, soldiers and Marines become better able to anticipate when they might find themselves in situations that require moral judgment and action. While the literal effects of actions may remain impossible to anticipate, recognition of the human costs of different military acts allows for a more comprehensive evaluation of the potential costs, not just the obvious costs to combatant and noncombatant lives and property. This improves the chances for victory and helps delineate what actions would constitute abuse. What is more, building this capacity within individual soldiers and Marines helps strengthen their resilience when they do face ambiguous or morally impossible situations in combat, reducing the psychological trauma they will suffer.

Conclusion: Moral Capacity Building for Irregular Warfare

Some resist the argument that it is possible to develop a soldier or Marine's capacity for moral behavior; by the time a person reaches adulthood, the argument goes, his moral compass is set and the soldier is either a "good egg" or a "bad egg." All that can be hoped for the "bad egg" is that the fear of punishment will serve as a sufficient restraint on his natural inclination for mischief. While a relatively common assumption, it is not born out in fact; rather, substantial empirical evidence exists to suggest that people are *more* receptive to moral training during adulthood. When it comes to developing moral sensitivity, the results of training are most pronounced among those who have not completed an undergraduate degree (those with a college education tending to have a higher pre-existing baseline) (Myyry and Helkama 2002). This means that there is the greatest possibility to develop the type of moral sensitivity needed in irregular warfare in enlisted soldiers and Marines—precisely the individuals who are conducting the street patrols and missions that require it. This section builds on Rest's model to outline what a training program might look like to better prepare soldiers and Marines to face the moral challenges of irregular warfare.

The first step is to provide the individual with positive cases that illustrate others at his rank demonstrating effective sensitivity. By starting with relatively easy cases (cases with minimal threat to the soldier or Marine) and increasing the degree of difficulty over time—by making the situation more ambiguous or elevating the risk of harm to the various actors given different choices—the individual soldier and Marine can cultivate the skill of discerning how different actions may affect the relevant actors in his area of operations (Robinson 2007, 33–4).

The Army's Center of Excellence on the Professional Military Ethic and the U.S. Naval Academy have developed on-line simulations that include precisely this type of training. While the programs are relatively new, it will be interesting to see how this exposure to moral sensitivity training influences junior soldiers' and Marines' ability to operate in complex environments.

An obvious, but critical point here is that while the scenarios are relatively simplistic at the beginning, they must grow more complex to be useful (Haidt and Joseph 2004, 65). If they remain simplistic, and therefore unrealistic, they fail to prepare the troops for the reality of what they will experience in irregular warfare. While no training can perfectly approximate combat, the closer to reality the scenarios become, the greater the conditioning effect the exercise has on building the muscle memory that will provide intuitive judgment in combat. This is another reason why positive cases are important. The patterns of moral action included in the scenarios will shape the troops' cognitive and affective processing of similar scenarios. If junior troops train only on ethical failures, they will lack the positive patterns to draw from when deployed.

Likewise, if they train and deploy in the absence of a small group leader who possesses the authority to model appropriate behavior, their ability to form the muscle memory needed to intuitively select moral judgments will be impeded. It is critical that the erosion of small unit cohesion that has resulted from the rapid and extended deployments be remedied if junior soldiers and Marines are to develop the moral capacity they need to fulfill their responsibility in irregular warfare. Someone with sufficient authority must be present to reinforce positive moral behavior prior to deployment (to begin building muscle memory) and as those patterns of behavior are challenged through stress and trauma. The small unit leaders are the ones who form and hold the morality of the units as a whole by setting and maintaining the command climate. They simply cannot do this if they do not bond with their units prior to deployment or if their units are cobbled together and pulled apart piecemeal throughout their deployment.

In terms of motivation, it is important to focus on developing soldiers' and Marines' internal motivation to behave morally, though the Army and Marine Corps must be careful institutionally not to contradict these efforts with external incentives that motivate behavior inconsistent with the restraint needed in irregular warfare (McNeal, 2009, 133). The Combat Action Ribbon and Badge have already been mentioned as one illustration of this; perhaps a different ribbon or badge could be awarded for those who effect the greatest change in their area of responsibility during a deployment. Likewise, as promotions reflect that those commanders who demonstrated moral behavior appropriate to irregular warfare are rewarded for their actions (and those who adopted an overly aggressive approach when inappropriate are not), younger troops will see an institution that supports the type of complex moral conduct that is required in irregular warfare.

All of these recommendations require continued effort to develop the character of those who serve in uniform. It is essential that our soldiers and Marines possess and display sound character; it is also essential they possess and display the other facets of moral action. By working to develop moral sensitivity, judgment, and motivation among junior soldiers and Marines, they become better prepared to wage irregular warfare effectively and well, with a reduced propensity for abuse and individual trauma. To the extent the United States is committed to waging this type of war, efforts to develop all four elements of moral behavior will pay significant dividends not only in our success at war, but in our success at supporting those who fight it on our behalf.

Bibliography

Aquino, K. and Reed, A. (2002), "The Self-Importance of Moral Identity," *Journal of Personality and Social Psychology*, 83:6, 1423–40.

Audi, R. (2008), "Intuition, Inference, and Rational Disagreement in Ethics," *Ethical Theory and Moral Practice*, 11:5, 475–92.

Bandura, A. (1999), "Moral Disengagement in the Perpetuation of Inhumanities," *Personality and Social Psychology Review*, 3:3, 193–209.

Bartone, P. (2004), "Understanding Prisoner Abuse at Abu Ghraib: Psychological Considerations and Leadership Implications," *The Military Psychologist*, 20:2, 12–16.

Breivik, G. (1999), "Empirical Studies of Risk Sports," *Selected Works*. Vol. 5 (Oslo: Norwegian School of Sport Sciences).

Forsythe, G., Snook, S., Lewis, P. and Bartone, P. (2005), "Professional Identity Development for 21st Century Army Officers," in D. Snider and L. Matthews (eds) *The Future of the Army Profession* (New York: McGraw Custom Publishing), 189–210.

Fine, C. (2006), "Is the Emotional Dog Wagging its Rational Tail, or Chasing it?: Reason in Moral Judgment," *Philosophical Explorations*, 9:1, 83–98.

French, S. (2003), *The Code of the Warrior: Exploring Warrior Values Past and Present* (Lanham, MD: Rowman and Littlefield Publishers).

Haidt, J. and Joseph, C. (2004), "Intuitive Ethics: How Innately Prepared Intuitions Generate Culturally Variable Virtues," *Daedalus,* 133:4, 55–66.

Hall, R.H. (1975), *Occupations and the Social Structure, 2nd Edn* (Englewood Cliffs, NJ: Prentice-Hall).

International Security Assistance Force (2009), *ISAF Tactical Directive* (unclassified), released July 6, 2009. www.nato.int/isaf/docu/official.../ Tactical_Directive_090706.pdf, September 29, 2009.

Jordan, J. (2007), "Taking the First Step Toward a Moral Action: A Review of Moral Sensitivity Measurement Across Domains," *The Journal of Genetic Psychology* 168:3, 323–59.

Kegan, R. (1998), *In Over Our Heads* (Cambridge, MA: Harvard University Press).

Kennett, J. and Fine, C. (2009), "Will the Real Moral Judgment Please Stand Up? The Implications of Social Intuitionist Models of Cognition for Meta-ethics and Moral Psychology," *Ethical Theory and Moral Practice*, 12:1, 77–96.

Krulak, C. (1999), "The Strategic Corporal: Leadership in the Three Block War," *Marines Magazine* (January).

Mattox, J.M. (2005), "The Moral Foundations of Army Officership," in D. Snider and L. Matthews (eds), *The Future of the Army Profession* (Boston: McGraw Hill Custom Publishing).

Mattis, J. and Hoffman, F. (2005), "Future Warfare: The Rise of Hybrid War," *Proceedings*.

McNeal, G. (2009), "Organizational Culture, Professional Ethics and Guantánamo," *Case Western Reserve Journal of International Law*, 42:1, 125–49.

Mental Health Advisory Team (MHAT) IV (2006), *Final Report. Office of the Surgeon, Multinational Force Iraq and Office of the Surgeon General* (United States Army Medical Command).

Mental Health Advisory Team (MHAT) V (2008), *Final Report. Office of the Surgeon, Multinational Force Iraq and Office of the Surgeon General* (United States Army Medical Command).

Milgram, S. (1974), *Obedience to Authority: An Experimental View* (New York: Harper and Row).

Muñoz-Rojas, D. and J-J. Frésard (2004), *The Roots of Behaviour in War: Understanding and Preventing IHL Violations* (Geneva: International Committee of the Red Cross).

Myyry, L. and Helkama, K. (2002), "The Role of Value Priorities and Professional Ethics Training in Moral Sensitivity," *Journal of Moral Education*, 31:1, 35–50.

Osiel, M. (1999), *Obeying Orders: Atrocity, Military Discipline and the Law of War* (New Brunswick, Transaction Publishers).

Petrie, A. (1967), *Individuality in Pain and Suffering* (Chicago: Chicago University Press).

Rest, J. (1994), "Background: Theory and Research," in J. Rest and D. Narvácz, *Moral Development in the Professions: Psychology and Applied Ethics* (Hillsdale, NJ: Lawrence Erlbaum Associates).

Richardson, R., Verweij, D. and Winslow, D. (2004), "Moral Fitness for Peace Operations," *Journal of Political and Military Sociology*, 32:1, 99–112.

Robinson, P. (2007), "Ethics Training and Development in the Military," *Parameters*, 37:1, 23–36.

Rosen, S.P. (2005), *War and Human Nature* (Princeton: Princeton University Press).

Silver, S. (2006), "Ethics and Combat: Thoughts for Small Unit Leaders," *Marine Corps Gazette*, November, 76–8.

Stanley, E., Getsinger, C. Spitaletta, J. and Jha, A. (2010), "Mind Fitness in Counterinsurgency: Addressing the Cognitive Requirements of Population-Centric Operations," *Military Review*, 90:1.

Taguba, A. (2004), *AR 15-6 Investigation of the 800th Military Police Brigade. Headquarters*, Department of the Army. Declassified by USCENTCOM 15 October, 2004.

Tyler, T., Callahan, P. and Frost, J. (2007), "Armed and Dangerous (?): Motivating Rule Adherence Among Agents of Social Control," *Law and Society Review*, 41:2, 457–92.

Wong, L. (2004), *Developing Adaptive Leaders: The Crucible Experience of Operation Iraqi Freedom* (Carlisle, PA: Strategic Studies Institute).

Zimbardo, P. (2008), *The Lucifer Effect: Understanding How Good People Turn Evil* (New York: Random House).
Zuckerman, M. (1979), *Sensation Seeking. Beyond the Optimal Level of Arousal* (Hillsdale: Lawrence Erlbaum Associates).

Chapter 14

Loyalty and Professionalization in the Military

Peter Olsthoorn

Introduction

After World War II, the Netherlands made an ill-conceived attempt to keep the Dutch East Indies as a colony. The Indies, having been in possession of the Dutch for some centuries, had been occupied by the Japanese during the war, and, somewhat unexpectedly for the Dutch, part of the population proved unwilling to succumb to Dutch rule again after Japan's capitulation, striving for an Indonesian Republic instead. For over four years, under the pretence of restoring order, the Dutch fought a counter insurgency campaign to no avail that gained international disapproval yet was supported by and large by the Dutch population and most of its politicians. The methods used were brutal, including hundreds of summary executions, the torture of prisoners, and the burning of villages, reminding some observers of the Nazi brutalities the Dutch had suffered in the preceding years. Poncke Princen, a Dutch soldier who had seen quite a few Nazi prisons from the inside in the years 1944–45, did not want to be part of what were euphemistically called "police actions." Aged 22, he deserted in 1948, as did about 26 others, though Princen was the only deserter to join the guerrillas that fought for independence he believed they were entitled to, as most of the world did. As a guerilla, he at least on one occasion fired at his former colleagues, possibly killing some of them. In 1949 Dutch soldiers, in an attempt to catch Princen, killed his first wife in their house while Princen made a narrow escape. After independence, Princen became an Indonesian national, yet subsequently spent eight years in prison for his fight for human rights in the then independent country.

The excesses in the former Dutch East Indies, meanwhile, were not talked about in the Netherlands in the years following the war in its former colony. This changed when in 1969 Dr. J.E. Hueting, a former conscript, went public with the fact that Dutch military personnel—including himself—had committed serious war crimes. He was called a traitor and worse. The Netherland's biggest newspaper called, in a headline, a performance by him on television "sickening." In the end, Hueting had to go into hiding. The subsequent debate in parliament did lead to the setting up of a government committee, resulting in a report that described about 600 excesses committed by the Dutch. Quite surprisingly, the committee itself concluded the report was seriously incomplete at best. Yet, despite all the turmoil, in the 1970s it was still possible for Princen to visit the Netherlands without attracting too much attention. In the nineties, however, with most of the veterans having ended their working lives and having plenty of time to think, a second visit met with great resistance with Princen receiving numerous death threats. The Dutch police actions, Princen's desertion, and the loss of the colony still proved to be a raw nerve. That the Indonesian struggle for independence had been legitimate was, however, no longer disputed by that time; what Princen was blamed for was his, in the veterans' view, flagrant disloyalty to his country and, especially, his comrades.

The harsh treatment that befell Princen—and Hueting—points to a special feature of disloyalty: the disloyal person is not only condemned because of the damage he causes, but also, and probably more so, for the disloyalty itself (see also Keller, 2007: 216). It is insult added to injury. Conscientious objectors and whistleblowers, for instance, are often penalized heavier than would seem proportional considering the harm they did. Possibly, this is the case because to be lacking in loyalty is seen as a serious character flaw. But is it?

Is Loyalty a Virtue?

Loyalty is not among the cardinal virtues as found in Plato's work, nor is it present in the more elaborate system of virtues of Thomas Aquinas. Still, loyalty is, of course, not a "new" virtue (just think of God's testing of Abraham, or Sophocles' *Antigone*), yet it is only since fairly recently that it goes under its own name. Unfortunately, the existing literature on loyalty is scant, and in the case of loyalty in the military bordering on the absent (exceptions are Wheeler 1973; Kaurin 2006; Snider 2008). This might be for a reason, for loyalty is a virtue that is rather hard to define. And the more one tries to grasp its content, the more questionable it becomes whether it really always is a virtue. It is perhaps so in the private sphere—being loyal to your spouse, for instance—yet with friends the situation already becomes somewhat more complex. How far, for example, should one go in protecting a friend who one knows is in the wrong? Outside that intimate sphere things are even less clear. For instance, being loyal to king, country, or organization when everything is well seems a bit easy, even gratuitous. It does not require a lot of virtue, and for that reason it cannot be that that is all there is to loyalty. On the other hand, remaining loyal in difficult times is not praiseworthy per se either. For example, standing behind fellow countrymen, colleagues, or organization when it is clear that they are at fault seems a highly undesirable form of loyalty, and certainly not virtuous. "Our country, right or wrong" cannot be right from a moral point of view (Primoratz, 2008, 208). It is this unquestioning patriotism that Princen lacked.

Yet, it seems that to a certain extent that is what loyalty is sometimes about: standing up for your group just because that is what it is, your group. In general, loyalty tends to signify in some way prioritizing the interests of an individual, a group, or a country, even when reason dictates a different direction (Ewin, 1992, 406; see for a critique of the view that loyalty consists of the prioritizing of interests Keller, 2007: 8–11). Loyalty, thus defined, "requires us to suspend our own independent judgement about its object" and "affects one's views of who merits what" (Ewin, 1992: 403, 406, 411). Some might argue at this point that misplaced loyalty is not loyalty at all, but that seems too easy a way out (see also ibid: 404). In fact, one might even say that the opposite is more likely to hold true: someone who is careful with his loyalties, weighing them cautiously against other values, is probably not someone we would describe as having loyalty as a paramount attribute (Keller, 2007: 158; see also Ewin, 1992: 411). So, if a virtue is a corrective to a temptation (Foot, 1997), then loyalty clearly does not always meet that criterion. Loyalty is probably a disposition, but whether it is a beneficial one remains to be seen. It has therefore been called a "gray virtue," one that can serve both good and bad causes alike (Miller, 2000: 8). But maybe this is even being too kind to loyalty; the fact that one can coherently speak of "bad" or "misplaced loyalty," while it is more or less nonsensical to talk about "bad" or "misplaced" justice, might be seen as an indication that it not a virtue in the first place (compare Ewin, 1992: 415, Keller, 2007).[1]

Is, by the same token, disloyalty perhaps not in all instances a vice? The whistleblower, for instance, is apparently acting disloyally but, provided he has the right reasons, clearly justified in doing so (ibid: 204, 215). We deem him for that reason, if he indeed acts on good grounds, not a disloyal person, i.e., not as someone willing, or even tending, to betray, deceive, desert, or let down. It seems that having a disloyal character cannot be seen as anything else but a severe defect. However, can it consistently be held that acting disloyally is not always a vice, when being a disloyal person clearly is a bad thing? Is the whistleblower's conduct disloyal in the first place?

Possibly, some of the confusion here may disappear to some extent by not seeing loyalty and disloyalty as opposites on the same dimension, as two excesses we have to find a mean between, but as two separate dimensions. The opposite of acting loyal (and being a loyal person) is the absence of loyalty in someone's deeds and character, i.e., the absence of the tendency to suspend judgement and to side with someone or something more or less unquestioningly. Someone never led by loyalty might be a strange creature, but not per se morally flawed. Likewise, the opposite of disloyalty is not loyalty but not being disloyal, that is not betraying the persons who trust you. Seen in that light, the whistleblower is not being disloyal, but someone who is not led (astray) by loyalty, in this case to his employer. The same line of reasoning might also hold true for conscientious objectors, and maybe even for Mr Princen. His case does illustrate how difficult it is to draw the distinctions; one could say that his refusal to take part in the police actions, and

---

1 Perhaps, one could speak coherently of misplaced justice, but one who does so clearly does not consider it to be, in fact, justice. Misplaced loyalty is, however, still loyalty.

thus his desertion, testify of a healthy lack of loyalty, but did his shooting at former colleagues not amount to disloyalty?

Loyalty and Justice

Following a different line of argument someone could maintain, however, that whistleblowers, conscientious objectors, and Princen's desertion (leaving the shooting aside for the moment) are not only not disloyal, yet may even qualify as loyal, albeit to a principle instead of to group or organization—Princen's autobiography (1995), for instance, points in that direction. Although often treated under one heading, one could wonder, though, if loyalty to a person, group, or nation on the one hand, and loyalty to a principle on the other, are really two manifestations of one phenomenon, or two different things altogether. Suspension of independent judgment, or the "willingness *not* to follow good judgment" (Ewin 1992: 412), is, in general, not required by loyalty to principle, to name one important difference. What is certain, however, is that both forms, group loyalty and loyalty to principle, often collide. The latter form we could call, because it often boils down to that, justice, and its demands often go against the claims made upon a person by group loyalty.

One could say that the inevitable clashes between both forms of loyalty are in fact conflicts between reason and emotion, on the assumption that loyalty to persons, groups, organizations, or countries is based on sentiment, while justice is founded on reason (Rorty 1997). Loyalty "is the sort of thing that one grows into rather than decides to have" (Ewin 1992: 408).[2] Some authors, most notably Michael Walzer, have pleaded that these group loyalties represent forms of "thick morality," and that those are not only stronger than, but are also prior to, forms of "thin morality" such as justice. It is universalistic concepts such as justice that are derived from particularistic loyalties, not the other way around. Thin moralities are impersonal in that they do not serve someone's particular interest, and bear no mark of their particular origin (1994: 7). Consequently, they are much feebler. Justice comes much harder than loyalty to a (small) group of people we (can) identify with. In difficult circumstances, this is all the more so: "The tougher things get, the more ties of loyalty to those near at hand tighten and those to everyone else slacken" (Rorty 1997: 139). Not surprisingly, Poncke Princen's former comrades felt utterly betrayed. What is exceptional is that Princen almost as a matter of fact preferred the "thin morality" of justice, or loyalty to principle, to loyalty to concrete groups or persons—perhaps somewhat tellingly, he proved to be quite lacking in spousal fidelity.

---

2   That sharp distinction between reason and emotion gets somewhat blurred if we follow Richard Rorty in his view that justice is nothing more than loyalty to a very large group: that of all human beings (1997).

Rather distressingly, accepting the notion of thick morality would mean that the extent of a soldier's moral obligation to the local population would depend on to what degree he sees that local population as "insiders." At first sight, this may look like a highly unsatisfactory conclusion. However, if and when military personnel actually has a moral duty to, for instance, run a higher risk when reducing the risks for the local population, is in fact not that easy to say. The question to what degree in such cases Western militaries have such an obligation to do more to protect civilians, even if this may cause more casualties at their own side, is in effect the broader question how far our obligations to total strangers go. Walzer clearly thinks this obligation goes rather far; "if saving civilian lives means risking soldiers' lives, the risk must be accepted," within due limits, of course (1992: 156; see for a critique regarding the due limits clause Shaw, 2005: 135). Interestingly, the two examples Walzer provides us with are both about soldiers trying to minimize the risks to their own occupied populations during World War II (1992: 157). Uncertain is to what degree there is a willingness on the part of military personnel to run the same risks protecting "outsiders." Notwithstanding his (later) views about the strength of thick moralities, Walzer thinks soldiers ought to have this readiness (ibid: 158). Not an inconsistency: just war theory lies within the domain of thin morality (Orend, 2000: 32).

Now, one could simply argue, and this is generally done from a consequentialist perspective, that one person indeed has far reaching obligations to another (Godwin, 1793; Parfit, 1987; Singer, 1972). As we saw in the above, loyalty is prioritizing the interests of some group or persons, and it is this prioritizing that is considered morally dubious (Parfit, 1987: 98). For a consequentialist, the idea of thick morality, a form of what a consequentialist would call common-sense morality, is morally erroneous because it cannot be consistently held that we have special obligations to those that are close to us that override our obligations to strangers. It is self-defeating because if everyone would prioritize the interests of special others above those of strangers, we all end up worse of (ibid: 95–108, 444). From a consequentialist's point of view, the fact that most people are more inclined to help their next of kin than unknown persons in faraway countries, even when the latter's peril is much greater, is therefore, although perhaps understandable and natural, not moral.

Clearly, consequentialism is here harking back to the utilitarian dogma that everyone's life and happiness are of equal weight. That was a novel thought when formulated in the eighteenth century, and a necessary antidote to class justice and the like. It was, essentially, a plea for thinner loyalties. By now, this notion has become commonplace to the extent that we do not even see the revolutionary character of it any longer. That is, as far as we are dealing with fellow countrymen. In our dealings with "outsiders," it is still rather novel. Militaries, for instance, are not in the business of promoting the greatest happiness to the greatest numbers irrespective of by whom those numbers are made up of, but in the business of to some extent preventing mishaps befalling their nearest and dearest. Militaries are not neutral as to whose lives they are risking or saving; they act from agent relative reasons, not from agent neutral reasons, meaning that the relationship in which the subject (stranger or colleague) stands to them, matters (ibid: 27). However, if we do not agree to the belief that every individual counts as much as another also in our dealings with non-westerners, the question then is what the ratio should be like (see for an attempt to answer that question Gerhard Øverland's chapter in this volume).

Yet, even if we acknowledge that we have such far reaching obligations to strangers, the problem remains that most people do not *feel* they have those obligations (and, as said, loyalty lies more within the domain of sentiment than of reason) let alone act on it. The predicament facing most militaries is therefore clear: while they are themselves organizations that put emphasis on, and thrive by, ties of loyalty, those who deploy troops are politicians who more and more further a universalistic, "thin" form of morality (see also Walzer 1994: 16), ironically enough often regarding conflicts that are a result of people allowing themselves to be led by thick loyalties. So, where the military works with rather narrow circles of loyalty because that is what military effectiveness on most accounts demands (see for example Wheeler 1973), politicians tend to take the widest possible circle as their starting point (see also Rorty 1997: 140). Human rights, and the fulfilment of basic needs, should be secured for everyone. In that respect every individual counts as much as another, yet in another respect he does not.

It has been noted by several authors, though probably foremost by Martin Shaw in his *The New Western Way of War*, that, when it comes to losses, civilian casualties among the local population are less important than Western military casualties (2005). It is perceived that way by both the militaries and the populations at large in the West, hence the emphasis not only on maximum force protection, but also on relatively safe ways of delivering firepower, such as artillery and bombers. Force protection sometimes seems to take precedence over the safety of the local population. This is, of course, nothing out of the ordinary and is in line with the findings of those who have pointed to the strength of "thick moralities." Minimizing the risks for Western soldiers in ways that increase the chances of civilian casualties among the local population stands, however, in rather sharp contrast with the universalistic ambitions behind most military interventions. Furthermore, although this risk-transfer will generally remain within the limits of the principle of double effect of the just war tradition in that civilian casualties are an unintended side-effect of legitimate attacks on military targets, it possibly falls short in the light of Michael Walzer's restatement of that principle, holding that the actor had to make efforts to avoid civilian casualties as much as possible, "accepting costs to himself" (1992: 155). So, although most authors are "in agreement that utilitarian ethics don't work well in the military setting" (Bonadonna 1994: 18) and despite Walzer's own convincing critique of the doctrine (1992), something might be said for a military ethics that takes into account the consequences for all parties involved.

Conflicting Loyalties

To give an example of the difficulty all this poses for Western militaries: sixty years after the "police actions," the Netherlands military is fighting another counter-insurgency in Asia, again to restore order, though this time under the scrutinizing eye of the media, and with a public that is sensitive to both the number of Dutch casualties and (albeit less so) the fate of the local population. These sensitivities clashed when on June 10, 2007 the district of Chora, in the Afghan province of Uruzgan, was surrounded by 300–1,000 Taliban fighters. At the time, the district had Dutch (about 60) and Afghan troops within its borders, and was also home to 75,000 Afghanis who depended on them for protection. Extra Afghan police forces were requested, yet the few reserves that were sent by the Minister of Interior to Uruzgan to help in the end refused to go to Chora. On June 16, the Dutch and Afghan troops came under attack.

Both journalists and military personnel who had been in the area later recounted having had associations with the Srebrenica debacle in July 1995. On the evening of July 11 1995, the day that the vastly superior, in numbers and weaponry, Serbian troops had taken the Muslim enclave, the Dutch Minister of Defence and his colleagues in the cabinet in their bunker in the Hague, at the time felt that the Dutch troops should show "solidarity" with the local population. In retrospect this sounds somewhat hollow, given the fact that, due to an insufficient mandate and a lack of men and weapons, the Dutch battalion had been unable to prevent either the fall of Srebrenica or the subsequent murder of 7,000 Muslims. The disaster stuck with the Dutch forces for a long time, even though a voluminous report by the Netherlands Institute for War Documentation (2002) years later would lay the responsibility at the doorstep of the government that had decided to send the troops; it had been blinded by a combination of universalistic humanitarian ambitions and the wish to enhance Dutch prestige abroad.

Something traumatic like that should not happen again. The decision was quickly made not to leave the local population in the hands of the Taliban, and large elements of the 500-men Dutch Battle Group were moved in over the next two days. Howitzers, Apaches and F-16s, not available in Srebrenica, were called in to assist the troops on the ground in Chora. In the end, control of the area was regained, and about 200 Taliban were killed. Next to this, an unknown number of civilians lost their lives. According to a report by the Afghanistan Independent Human Rights Commission and the United Nations Assistance Mission in Afghanistan (2007), estimates range from 30 to 88, with 60 to 70 being a more realistic estimate. Some of them had been tortured, shot or beheaded, and torched by the Taliban, yet there is no doubt that others died as a result of artillery bombardments by ISAF, despite efforts to warn the local population beforehand using loudspeakers.

According to the report the methods used by ISAF were heavy-handed, and not always accurate. President Karzai and the ISAF Commander, US general Dan McNeill, criticized the Dutch for using a Howitzer, positioned 30 kilometers from Chora, without a forward controller. According to McNeill, in a classified report, this last element was a breach of the law of war. According to others, and among them Dutch Secretary General of NATO De Hoop Scheffer, a modern howitzer can do without a forward controller. A potential debate on how far the Western militaries' moral obligations to the local population go, and when force protection becomes risk aversion, was thus reduced to a dispute on technical issues. Later, in October 2007, it became apparent that Australian troops in Uruzgan, operating under Dutch command, had refused to participate in the operation, worrying about the risks the operation would impose on civilians, and being of the opinion that it was not in accordance with the rules of engagement. The Australian misgivings notwithstanding, a Dutch district attorney decided in June 2008 not to prosecute military personnel for what happened in Chora because they had acted within the limits set by the law of war and their rules of engagement. During the battle for Chora, one Dutch sergeant-major died due to an accident with a mortar.

Interestingly, this criticism on the decisions taken by the Dutch military in the biggest battle by the Netherlands military since Korea came some months after critique on the Dutch approach that at first seemed to be of an opposite character. In December 2006 some high-ranking Canadian and British officers testified to seeing this approach as essentially flawed, because it avoided doing what is a precondition for rebuilding Afghanistan: dismantling the Taliban. According to them the relatively small numbers of Dutch casualties gave evidence to its misguided weariness, cowardice even (critique that, too, brought back to the minds of many in the Netherlands the Srebrenica debacle and the following—and ongoing—debate in the media whether the Dutch lack courage). In fact, both criticisms—too careful at first, too heavy-handed later—in effect come down to the same thing: Dutch military personnel are not willing enough to run risks at their own peril. Probably, similar criticisms could be launched against all NATO and US troops in Afghanistan: the number of civilian losses caused by airpower tripled from 2006 to 2007, mainly due to the increased use of air support called in by ground troops in unexpected contact with the Taliban (Human Rights Watch 2008: 14).

Now, even though one could perhaps say it is to be expected that the military places the lives of soldiers above that of the local population (see for instance Cook 2002: 347), it is also to some extent inconsistent with the intent behind missions like these, and the tactics employed: trying to prevent the Afghanis from developing loyalties to the Taliban. The rising civilian death toll is thought to increase support for, and facilitate recruitment by, the Taliban (Human Rights Watch 2008). As it stands, military codes, military oaths, value systems and culture, in short: the military ethic (but also the national political viewpoints and the popular sentiment) seem however more or less antagonistic to the idea that the life of a local civilian counts for the same as that of a Western soldier. This ethic took shape, though, at a time in which the local population played a lesser role, and a "thicker" form of loyalty (defence of one's own country) prevailed. The issue here is whether the military ethic can be at least *somewhat* reformulated as to include the interests of the local population more than currently is the case, and what such an ethic should look like to bring the military profession closer to the ethic of other professions, which in general gives priority to those at the receiving end.

Professionalization and the Focus of Loyalty

Contrary to what is published on loyalty, much has been written in the last few decades or so on the professionalization of armed forces (see, for instance, Wolfendale, 2007). Professionalization in the case of the military is often taken to mean something as making use of volunteers (instead of conscripts) who are skilled, well-educated, etc. James Burk, for instance, defines a profession as "a relatively 'high status' occupation whose members apply abstract knowledge to solve problems in a particular field of endeavor" (2002: 21). Now and then, a professional within the armed forces is described as someone who is prepared to loyally fulfil the missions imposed by politicians, leaving one's own opinions aside. This latter depiction, however, seems to be at odds with what is generally understood by the term professional: someone with an expert body of knowledge and, therefore, a considerable degree of discretion and professional autonomy (see also Cook 2002: 342–43).

There is, however, yet another side to professionalism. In general, a professional is someone who is loyal to his profession and his professional ethic, more than to his organization. In practice this loyalty to a professional ethic seems to somewhat resemble loyalty to principle, most importantly in that it, too, does not ask for the suspension of independent judgment. Quite the contrary, the standards of a profession are general, in the sense that they are not confined to the institution of a professional, and have largely originated in universities and professional associations, not in the organization he is employed by (Mintzberg 1983: 192). A surgeon, for example, is often more concerned about the judgements of other surgeons, such as those brought to him in the verdicts of his professional association, than those of his happen-to-be-employer of that moment. Furthermore, he tends to place his own professional judgement above that of the management that supervises his work, based on the idea that his training, education, and professional experience makes his verdict a more informed one compared to that of those above him in the organization. That his profession is the focus of his loyalty makes the professional someone who can relatively easily, without too much pain in the heart, switch from one employer to another. More importantly, the interests of colleagues or an organization have less room in professional ethics than those of a professional's clients.

It is this side of professionalism, loyalty to the profession instead of the organization, which is, not surprisingly, less developed in the military. Quite some effort is taken to ensure military personnel are loyal to their own organization; they are meticulously socialized into the armed forces, not into a profession. There are one's own values (and upholding these as required by the virtue of integrity—something that is at least paid lip service to by most militaries), and the organizational values, but there are, in the case of the military, not really any values of the profession. This socialization into the organization is made easier by the fact that, where doctors and lawyers receive most of their formal professional training before entering their job, military personnel is predominantly trained in house. Furthermore, professional associations, compared to similar associations in other professions, traditionally do not play much of a role in the development of the profession (and as a result do not enhance loyalty to the profession), as they mainly look after the material interests of their members (Van Doorn 1975: 36). Viewing some concrete examples—oaths, value and virtue lists, codes of conduct—to see how exactly this side of professionalism has taken shape in other professions, compared to the military, might be instructive, especially since unethical behavior is rarely a result of failings at the individual level; it is more often a result of the situation, including the ethical climate in the organization partly shaped by these oaths, codes of conducts, and value lists.

The Military Ethic and Professional Ethics

Most manifestations of military ethics and other professional ethics respectively, are fairly consistent as to whose interests are most important, though pointing in different directions. An example of a professional oath is the medical oath which comes in many varieties. The common denominator is that a doctor should work in the interest of his patients; there is no mentioning of parties outside the doctor-patient relationship, such as hospitals or governments. The military oath is different in this respect; it also comes in many forms, but it often stresses loyalty to a constitution or head of state. The people at the receiving end, such as the local population in Iraq or Afghanistan, are not mentioned. Interestingly, these two different oaths, medical and military, can—and do—lead to dual loyalties in the case of doctors working for the armed force (see for instance Clark, 2006). Similar dual loyalties might be experienced by other professionals in the military, for instance counselors, lawyers, or controllers.

Roughly the same pattern emerges when comparing the values of the military profession with those of other professions. Armed Forces are, first of all, less hesitant than most ethicists about loyalty's beneficial properties. They often include it in their lists of values (Robinson 2008) and do consider it a cardinal virtue. Yet the question remains, loyal to what? Now, as mentioned earlier, one can be loyal to a clan, tribe, or organization, but also, by most accounts morally on a higher plane, to a principle, an ethic, or a code. Apparently, as indicated earlier, loyalty to an organization is not only different from loyalty to principle; given the fact that most organizations go astray every so often, these two loyalties will even clash at times. The question which of both forms of loyalty should prevail if their demands are incompatible is easily answered for military personnel as far as the organization is concerned: clearly, it is loyalty to the organization that militaries stress, more than loyalty to principle, which is hardly ever mentioned in the various armed forces value statements. Country, colleagues, and the organization, are central in most descriptions. This is in line with how other values are seen: they, including respect, also mainly relate to colleagues. That does raise the question how loyalty to an organization is to be combined with other values, for instance the loyalty to principle which integrity requires, or the moral courage that is included under the heading of courage by most militaries. In effect, moral courage, like integrity, also comes very close to loyalty to principle.

Whereas in the value lists of the different armed forces military effectiveness and the interests of organizations and colleagues hold central place (Robinson, 2008), the values of the medical profession, formulated by professional organizations, give precedence to the patient and the doctor-patient relationship, stating for example that the practitioner should always work in the interest of the patient, refrain from prescribing treatment he knows is harmful, and respect the patient's dignity. Medical ethics is about patients and medical care, not about how to associate with colleagues. Somewhat similar, police forces, a bit more akin to the military profession, put the interests of civilians first in their value statements.

Lastly turning to codes of conduct, we see, for instance, how the furthering of international rule of law is mentioned in the Netherlands Constitution, but that the code of conduct of its armed forces stresses primarily, among other things, the importance of being a team member who maintains his professional skills and does not tolerate discrimination, sexual intimidation, and the like; treats everybody with respect, and contributes to a safe working environment. As such, it is more, although not exclusively, about regulating the conduct of military personnel towards each other than it is about their conduct towards those they are to protect. A look at the accompanying explanation shows that even the "everyone" in the sentence "I treat everyone with respect" appears to refer to colleagues and not to outsiders. Codes of conduct for doctors, to the contrary, do emphasize the interests of outsiders, namely the patients, as do those of the police. How worrying the omission of outsiders in military codes of conduct is, is uncertain; according to most recent literature, the influence of codes of conduct is limited anyway (see for an overview: Verweij, Hofhuis, and Soeters 2007: 19–40). Nevertheless, the gist of the codes, oaths and values as currently formulated in most militaries is undeniably somewhat one-sided in that they mainly pay attention to the needs of the organization and colleagues.

The clear difference between the medical and the military profession in this respect, and the conflicting loyalties that can result from that for military doctors, testify to the fact that a doctor in the military serves in his capacity as a physician a different client than in his role as a member of the armed forces. For a civilian doctor, on the other hand, it is manifest that it is the patient who is his client, and no one else. This seems to be characteristic of a professional, defined by one author as someone who "works relatively independently of his colleagues, but closely with the clients he serves" (Mintzberg 1983: 190). Disregarding working independently (although that element is absent in the military does pose the question to what

measure it is justified to speak of a military profession in the first place), for military
personnel the client, if we want to use that term, is the state or the people, not the
local population in Afghanistan, Iraq, or Sudan. It seems they are often not even
considered a party. The UK Army, for example, states that a "mutual obligation
forms the Military Covenant between the Nation, the Army and each individual
soldier; an unbreakable common bond of identity, loyalty and responsibility which
has sustained the Army throughout its history." Don Snider, writing about the US
Army, mentions "three critical trust relationships of the military profession—those
with the American people, those with civilian and military leaders at the highest
levels of decision-making, and those with the junior corps of officers and noncommissioned officers of our armed forces" (2008: vii). Given this strong and
exclusive emphasis on institutional loyalty, it is perhaps not surprising that, when
military doctors have to choose between their responsibility for their patients and
the demands of the military, they sometimes have their obligations to their patients
overridden by their sense of military duty (Clark, 2006: 577).

What is to be Done?

As we saw, loyalty does figure prominently in the lists of virtues and values of
most armed forces. Yet, these armed forces do not insist on loyalty to principle, nor
do they demand loyalty to professional ethics, which is, as said, in some aspects
somewhat similar to loyalty to principle. One could argue that there is some logic
in all this. To state the obvious: militaries cannot have too many soldiers taking
an impartial, "objective" view on matters like Poncke Princen did. Furthermore,
military organisations ask much of their personnel, such as running considerable
risks, which they cannot be expected to fulfil in return for a salary alone. As stated
above, loyalty is not just about standing behind an employer when all is well;
particularly military organisations are dependent on people who will stand their
ground in difficult circumstances. Indeed, the position of, for instance, a police
officer or a practitioner of medicine is, in essence, different from military personnel
deployed in Iraq or Afghanistan. A doctor can put the interest of his patients above
everything else without putting himself in harm's way, whereas the latter cannot.
Emphasizing loyalty to the organization is therefore not surprising considering the
predominant task of a defense organization is the defense of its own territory. In
the past, it has therefore been maintained that the military profession was ill-suited
to develop into a "true" profession (Van Doorn 1975).
    At the same time, what seems to be behind the many moral dilemmas military
personnel faces today is often a conflict between loyalty to a group—one's
team or organization—on the one hand and loyalty to principle on the other
hand. Interestingly, the moral exemplars frequently used in, for instance, ethics
education—such as Hugh Thompson Jr., the helicopter pilot who tried to stop the
My Lai massacre—chose loyalty to principle, i.e., justice, above loyalty to their
organization and colleagues. One might ask the question, therefore, if, at a time

that many armed forces consider the promotion of universal principles as their main ground for existence, the development of a true professionalism, with the main focus of loyalty being a professional ethic instead of one's organization, is still too far-fetched. Thompson, incidentally, suggested that there is a basis for such professionalism when he said that those involved in the killing in My Lai "were not soldiers. They were not military people," indicating that there can be a loyalty to a professional military ethic distinct from loyalty to the organization.[3] The fact that he in subsequent years was shunned and threatened by colleagues suggests, however, that this professional military ethic needs some strengthening.

To make a beginning with this, the most important measure to be taken possibly lies in somewhat adjusting the self-image, or social identity of military personnel (Franke, 2003). The recent wars to topple the regimes in Afghanistan and Iraq excluded, for most of the militaries in the Western world peacekeeping and humanitarian missions are becoming their core business. As it stands, these "Operations Other Than War" are sometimes seen as lesser than "the real thing." US Marines, for instance, follow a martial arts program nowadays, aimed at keeping them in touch with their warrior ethos after an era in which the military was seen "as an instrument of social engineering" (Yi, 2004). Research into the behavior of US military personnel participating in Operation Restore Hope in Somalia during the early nineties found, however, that military personnel falling back on warrior strategies during peacekeeping are more likely to have escalating contacts with the local population (Miller and Moskos, 1995). Oaths, codes of conduct, and value and virtue lists that give due place to the interests of outsiders can contribute to altering that self-image. As far as those value and virtue lists are concerned, there is no need to immediately replace the existing values by more cosmopolitan ones. Interpreting the current values, especially loyalty, somewhat more comprehensively is already an, admittedly small, step in the direction of a more professional ethic. Possibly, it could also include enemy soldiers, substituting controlled or even minimal use of force for overwhelming force. In addition, the increased deployments under the flag of NATO, EU, and the UN might, too, contribute to a more professional, and less organizational or national, ethic (see also Van Doorn 1975: 42). So far, however, these organizations have not done much in developing their own values and codes.

Parallel to the increase in joint operations, military academies are more and more conferring with fellow academies about their ethics programs, trying to learn something of how things are undertaken elsewhere during conferences and workshops. The Military Ethics Education Network founded in May 2008 (MEEN), for instance, not only aims at comparing and examining the values and virtues emphasized by Western militaries, assessing the various approaches and methods of ethical education adopted in the different institutions and gauging their effectiveness, but also at producing textbooks and case studies. The network might function as a professional organization equally found in other professions.

---

3   See http://www.usna.edu/ethics/Publications/ThompsonPg1-28_Final.pdf.

Finally, some form of protection for whistleblowers, not disloyal of character yet just not willing to suspend good judgment, is probably a good idea in every line of work, though seems especially needed in the military, for those who "tell" not just do find themselves out of a job. Abu Ghraib whistleblower Joe Darby feared for his life for a while, and lived for years at an undisclosed location, in protective custody.

Conclusion

A world without loyalty, with, for instance, parents taking an objective view to their children, or friends looking in a similar vain towards their friends, would be somewhat shallow. In that sense, loyalty might qualify as a value. However, as we have seen, it is most likely not a virtue per se. Nonetheless, the question is whether loyalty is a praiseworthy or even necessary quality in military personnel. It seems to be a quality more beneficial to one's colleagues than outsiders. On the other hand, if we see loyalty as, at least partly, consisting of the suspension of independent judgment, one could wonder whether it is in fact helpful for colleagues in the long run; that is, if we take loyalty as it is defined and demanded by the military itself: loyalty to the organization and colleagues. Loyalty to principles, not asking for the suspension of free judgment, is not promoted by militaries under the banner of loyalty (though probably present in the often mentioned value of integrity), yet can be expected to be beneficial to colleagues and outsiders alike. Now we would clearly be expecting the impossible if we wished militaries to substitute the one form of loyalty for the other, that is to say, loyalty to principle for group loyalty. However, there seem to exist some shades of gray between group loyalty and loyalty to principle. The German officers that plotted against Hitler, or some examples of whistleblowers, for instance, although clearly denouncing blind group loyalty, did not act from universal principles either. Hitler's adversary Von Staufenberg was motivated by loyalty, yet not to the Germany as it was, but to an idea of what Germany should be. Whistleblowers, likewise, although not acting how the organization desires, sometimes profess to act from values their organization once held in their opinion, yet alas has departed from (see also Hirschman 1970: 76–105). Hugh Thompson seems to be a case in point.

These examples suggest that a middle ground between loyalty to principle and group loyalty is possible. The development of a more professional ethic in the proper sense of that term, with the focus on loyalty to a profession rather than an organization, would be holding that middle ground and is therefore perhaps an attainable ideal. In a sense, it would also be holding the middle ground between the hard to attain universalism of the consequentialist and the sometimes somewhat slippery particularism of Walzer. Although not as principled as loyalty to principle, loyalty to a professional ethic would be a lot more so than the current organizational and group loyalties. If there is any truth in the often heard suggestion that tactics that are relatively safe for military personnel yet put civilians at risk in fact can drive the local population into the hands of the insurgents, such a more professional ethic will benefit outsiders as well as colleagues and the organization in the long run.

References

Afghan Independent Human Rights Commission and the United Nations Assistance Mission in Afghanistan (2007), *IHRC and UNAMA Joint Investigation into the Civilian Deaths Caused by the ISAF Operation in response to a Taliban Attack in Chora District, Uruzgan, on 16th June 2007.*

Bonadonna, R.R. (1994), "Above and Beyond: Marines and Virtue Ethics," *Marine Corps Gazette*, 78:1.

Burk, J. (2002), "Expertise, Jurisdiction, and Legitimacy of the Military Profession," in D.M. Snider and G.L. Watkins (eds), *The Future of the Army Profession* (New York: McGraw-Hill Primus Custom Publishing), 19–38.

Clark, P.A. (2006), "Medical Ethics at Guantánamo Bay and Abu Ghraib: The Problem of Dual Loyalty," *The Journal of Law, Medicine & Ethics*, 34:3, 570–80.

Cook M.L. (2002), "Army Professionalism: Service to What Ends?," in D.M. Snider and G.L. Watkins (eds), *The Future of the Army Profession* (New York: McGraw-Hill Primus Custom Publishing), 337–54.

Doorn, J. van (1975), *The Soldier and Social Change* (London: Sage Publications).

Ewin, R.E. (1992), "Loyalty and Virtues," *Philosophical Quarterly*, 42:169, 403–19.

Foot, P. (1978), *Virtue Ethics* (Oxford: Oxford University Press).

Franke, V. (2003), "The Social Identity of Peacekeeping," in T.W. Britt and A.B. Adler (eds), *The Psychology of the Peacekeeper: Lessons from the Field* (Westport, CT: Praeger), 31–52.

Godwin, W. (1793), *An Enquiry Concerning Political Justice, and Its Influence on General Virtue and Happiness* (London: G.G.J. and J. Robinson).

Hirschman, A.O. (1970), *Exit, Voice, and Loyalty: Responses to Decline in Firms, Organizations, and States* (Cambridge, MA: Harvard University Press).

Human Rights Watch (2008), *"Troops in Contact" Airstrikes and Civilian Deaths in Afghanistan*, http://hrw.org/reports/2008/afghanistan0908/.

Kaurin, P. (2006), "Identity, Loyalty and Combat Effectiveness. A Cautionary Tale," JSCOPE paper, http://www.usafa.edu/isme/JSCOPE06/Kaurin06.html.

Keller, S. (2007), *The Limits of Loyalty* (Cambridge: Cambridge University Press).

Miller, L.L. and C. Moskos (1995), "Humanitarians or Warriors? Race, Gender, and Combat Status in Operation Restore Hope," *Armed Forces and Society*, 21:4, 615–37.

Mintzberg, H. (1983), *Structure in Fives* (Englewood Cliffs, NJ: Prentice-Hall).

Netherlands Institute for War Documentation (NIOD) (2002), *Srebrenica—A "Safe" Area. Reconstruction, Background: Consequences and Analyses of the Fall of a Safe Area.*

Orend, B. (2000), *Michael Walzer on War and Justice* (Cardiff: University of Wales Press).

Parfit, D. (1987), *Reasons and Persons* (Oxford: Oxford University Press).

Primoratz, I. (2008), "Patriotism and Morality: Mapping the Terrain," *Journal of Moral Philosophy*, 5:2, 204–26.

Princen, P. (1995), *Een kwestie van kiezen* (Den Haag: Uitgeverij: BZZTôH).

Robinson, P. (2008), "Introduction: Ethics Education in the Military," in P. Robinson, N. De Lee and D. Carrick (eds), *Ethics Education in the Military* (Aldershot: Ashgate).

Rorty, R. (1997), "Justice as a Larger Loyalty," *Ethical Perspectives*, 4:3.

Shaw, M. (2005), *The New Western Way of War* (Cambridge: Polity Press).

Singer, P. (1972), "Famine, Affluence, and Morality," *Philosophy and Public Affairs*, 1:1, 229–43.

Snider, D.M. (2008), *Dissent and Strategic Leadership of the Military Professions* (Carlisle Barracks, PA: Strategic Studies Institute, U.S. Army War College).

Yi, J. (2004), "MCMAP and the Warrior Ethos," *Military Review*, November–December.

Walzer, M. (1992), *Just and Unjust Wars* (New York: Basic Books).

Walzer, M. (1994), *Thick and Thin: Moral Argument at Home and Abroad* (Notre Dame, IN: Notre Dame Press).

Wheeler, M.O. (1973), "Loyalty, Honor, and the Modern Military," *Air University Review*, May–June.

Wolfendale, J. (2007), *Torture and the Military Profession* (Basingstoke, UK: Palgrave-Macmillan).

Index

Bold page numbers indicate figures, *italic* numbers indicate tables.